Leo Hartley Grindon

The Shakspere Flora

Leo Hartley Grindon

The Shakspere Flora

ISBN/EAN: 9783337377038

Printed in Europe, USA, Canada, Australia, Japan

Cover: Foto ©berggeist007 / pixelio.de

More available books at **www.hansebooks.com**

THE
SHAKSPERE FLORA.

A GUIDE TO ALL THE PRINCIPAL PASSAGES IN WHICH
MENTION IS MADE OF

Trees, Plants, Flowers, and Vegetable
Productions;

WITH COMMENTS AND BOTANICAL PARTICULARS.

BY LEO H. GRINDON,

*Author of "Lancashire: Historical and Descriptive;" "Country
Rambles;" "Manchester Banks and Bankers;" "The
Little Things of Nature;" and other works.*

Then to the well-trod stage anon,
If Jonson's learned sack be on,
Or sweetest Shakspere, fancy's child,
Warble his native wood-notes wild.
L' Allegro.

MANCHESTER:
PALMER & HOWE, 73, 75, AND 77, PRINCESS–ST.
LONDON: SIMPKIN, MARSHALL, & CO.

1883.

PREFACE.

O results of an acquaintance with Shakspere are more useful, pleasing, and varied, than such as arise upon familiarity with his references to trees and plants. These references are key-notes to far more than appears upon the surface:—in the following pages it is attempted to show to what they lead; to bring out the poetry of every allusion, where poetry is involved; and to show the purpose and significance of the terms employed. Every tree, plant, flower, and vegetable production

named in the plays and poems, is dealt with; every important passage in which the name occurs, is quoted;—the characters speaking are also dealt with, when the beauty of the reference makes this desirable.

That the practised Shaksperean critic will in this volume find anything new, I do not for a moment presume to suppose. Happily there is now a daily growing national interest in Shakspere, as shown particularly by the increase, during the last few years, of Shakspere Societies, and the introduction of Shakspere as one of the chief levers of education in all high-class schools. This encourages me to hope that I am at all events providing a guide to Shakspere's plants which may be useful to beginners, and to these I offer such help as they may find. For the sake also of students, endeavour is made to explain any curious and unusual words occurring in the passages quoted.

So far as regards my predecessors in Shaksperean commentary, I ask only to be regarded as a technical botanist, following in their wake, and possibly suggesting an illustration here and there that may have been overlooked.

"Shakspere" is adopted because the shortest of the various spellings of the name, and because approved by the New Shakspere Society of London,—if erring with whom, I am at least in good company. Incomparably more important than settling which of the several modes may be the right one, is the striving with all our power to win over as many as we can of the rising generation to reverence for the poet himself.

On page 17, line 7 from bottom, "birch" has been accidentally printed instead of "larch."

LEO H. GRINDON.

Manchester, May, 1883.

CONTENTS.

———

CHAPTER V.

CHAPTER VI.

CHAPTER VII.

CHAPTER VIII.

CHAPTER IX.

FULL-PAGE ILLUSTRATIONS.

*We are indebted to Mr. Thos. Letherbrow for the Drawing
and Etching of the above Illustrations.*

The Shakspere Flora.

Chapter First.

INTRODUCTORY.

How sweet the moonlight sleeps upon this bank !
Here will we sit, and let the sounds of music
Creep in our ears.—*Merchant of Venice*, v., 1.

HAKSPERE, pre-eminently the poet of the heart—no man having unveiled as he has done, the many-hued mystery of human emotion—is no less the poet, supremely, of Nature. He sees everything, both great and small, which environs us, and this not only with the eye of the artist, intent upon loveliness of form and colour, but with the profounder comprehensiveness which gives ability to interpret, asking what may these things signify, what may be their story for the

B

imagination.   He is not slow either to observe the
phenomena of which science takes special cognizance.
Were he indifferent to them, he would not in truth be
Shakspere, nor even Poet.   The number of simple facts
belonging to the "scientific" class referred to by Shakspere is in itself something wonderful.   His allusions go
far also towards illustrating the ancient and uncontested
doctrine that every great discovery in science is the
fulfilment of a prophecy put forth by some noble poet of
an earlier age.

Shakspere was not a botanist; not a man skilled, that
is to say, in the technical knowledge of plants.   That he
should be one was by no means needful to his pre-
eminence as poet.   Had his attention been fastened
upon plants and flowers; had he qualified himself to
write a book about them, probably we should in other
respects have lost, and so far we may be thankful that he
looked upon them simply as most other men do, *i.e.*, as
charming factors in the general loveliness of nature,
which it is for individuals of special taste to study and
classify.   What Shakspere has done for us botanically,
and in this respect so excellently, is found in sweeter,
and more loving, and more copious mention of plants
and flowers than occurs in any other single writer, not
professedly technical, the world has seen.   For this we
may be thankful indeed.   It is quite enough to be glad
of; and in the consideration of his forms and methods a
little time is not ill spent, even by those who care
nothing at all for botany as a science.

Comparing Shakspere with other poetic literature, in respect of profusion of reference to trees and plants, he is surpassed only by the Old Testament writers. After him come two or three of the classical authors—Theocritus, in the Idylls, and Virgil, in the Eclogues and the Georgics; in later times, those we may call relatively our own, next in order stand Chaucer and Milton. Poets who laid themselves out especially for botanical themes — Erasmus Darwin for example — we do not count; the charm of all such allusions as are now referred to consisting in their purely spontaneous character. Shakspere's purpose, like Chaucer's, was to depict human action and human character in its varied phases, not to talk about wild-flowers. The mention of them, usually, is exquisite; the introduction of their names is still only incidental. Having them, we recognize their fitness of place, and rejoice in their presence; yet were they absent they would not be missed. In no case is it desirable or even prudent to draw any exact comparison as to the quality of the references, and least of all as regards Milton's, though the tendency is no doubt always to couple the names and to look for resemblances. Theological narratives seldom give seasonable opportunities for allusions other than the purely figurative and fanciful. In the entire mass of the two great Protestant epics, not forgetting the picture of the garden of Eden, there is not a passage to put abreast of Perdita's "O Proserpina!" In that most chaste and finished of elegies, *Lycidas*, there is scope enough, and in *Comus* room is not wanting,

and in these two poems Milton shows well. But we are
still thrown back upon the superiority of Shakspere in
respect of affectionate intimacy with the things around
him, as distinguished from the traditional and far-distant,
since a poet is always delightful in the degree that he
keeps alive our warmest and tenderest home-sympathies.
Milton, as a learned classical scholar, cites chiefly the
plants already famous in the literature of ancient Greece
and Rome. He talks of the amaranth, of Homer's moly,
of the acanthus, the asphodel, "cassia, nard, and balm."
Shakspere, on the other hand, excels in regard to the
flowers of the wayside, the pretty objects it needs no
classical knowledge to understand. Milton fails also in
respect of accuracy. His practical knowledge of the
things around him was far below Shakspere's, so that—
although Shakspere himself in one place makes a slip,
and although at times his own epithets are matchless, as
when in *Lycidas* we have the woodbine "well-attired"
—unfortunately there are many distinct mistakes. The
violet is not a "glowing" flower; wild-thyme does not
mingle with the "gadding vine;" nor does the jessamine
"rear high its flourished head." The sublimity of
Paradise Lost, the music and the incomparable grace of
Lycidas and *Comus*, are qualities of course altogether
independent of the botany, and would remain were all
the botany detached. These are not the present consideration, which is simply the inexpediency of comparing
where similarities are in the very nature of things few
and distant.

The superiority in point of numbers in the Shaksperean allusions may seem at the first glance to come of the vastly greater extent of surface of his writings. But this does not affect the question, since in at least five or six of the dramas there are not more than a dozen plant-allusions, and these very often meagre ones. Among the plays specially deficient in regard to number of references are *Julius Cæsar*, *Much Ado About Nothing*, and the *Two Gentlemen of Verona*. Very few occur either in the *Merchant of Venice*, in *Macbeth*, in *King John*, *King Richard the Third*, and *Measure for Measure*. They arrive in plenty, on the other hand, in the *Tempest*, *Love's Labour's Lost*, *Hamlet*, *Romeo and Juliet*, *King Lear*, *Cymbeline*, and still increasingly in the *Winter's Tale* and the *Midsummer Night's Dream*. Not that all are of corresponding poetic eloquence, or fraught with pleasing associations. Often there is no more than the barest mention, and the word seems to serve no purpose but that of completing the measure. In all ages the poets have been prone to introduce names of objects, and even epithets, for the sake purely of metre or euphony. To do so, when genius shines on every page, so far from being a sign of weakness, is in harmony with the wise deliberate repose never disdained or forgotten by Strength. The solitary Shaksperean botanical slip is, like all his other lapses, so palpable as to be detected on the instant. It occurs in the scene in *Cymbeline*, where Imogen, lying upon her bed, asleep and half disrobed, is contemplated by Iachimo. He notes her closed eyelids,

> White, and azure-laced
> With blue of heaven's own tinct;

then,

> On her left breast,
> A mole cinque-spotted, like the crimson drops
> I' the bottom of a cowslip.

A certain amount of latitude is always permissible in descriptions designed to be vivid and picturesque, but it is going quite beyond the reality to say that the spots in the cup of the cowslip are "crimson." The nearest approach to that colour ever seen could only be described as rosy orange. Considering the infinite number of printers' and copyists' errors in the early editions of Shakspere, the famous "first folio" of 1623 taking pre-cedence of all in point of inaccuracy;* considering, too, that in spite of all the criticism bestowed upon Shakspere we have not yet got the poet's exact words in many an obscure passage, nor even the general sense of the phrase—this word "crimson" might at first sight be thought one of the terms awaiting correction. But there is no reason to doubt the authenticity, and we must be content to take it as the exception to the author's other-wise unbroken faithfulness to nature. This, in truth, is the only circumstance which justifies or even calls for

---

* "Perhaps," says the Rev. Joseph Hunter, in the preface to his *New Illustrations of Shakspere*, "Perhaps in the whole annals of English typography there is no record of any book of any extent, and any reputation, having been dismissed from the press with less care and attention than the first folio of 1623" (the first collective edition).

the present comment. Never mind. We have no time
to spare for it. Set against this little slip,—

> Daffodils
> That come before the swallow dares, and take
> The winds of March with beauty,

or any one of a score of similar perfections, and it is a
thousand times outweighted. Whoever else you may be
disposed to quarrel with, never fall out with Shakspere;
not even over a verbal error, any more than, if you would
prosper, over a rule of life. Discover the misprints by
all means; amend everything that is clearly not Shakspere's own; but let the blemishes pass. In works of
noble art, before we attend to the flaws, it is no more
than common-sense to make sure that we first perceive
all the merits. Life is too short, opportunity too limited,
for learning even a small portion of the great and
glorious. Why, then, waste ever so little of either in
search for blots and weaknesses? There is no real skill
or cleverness in discovering the faulty. Far away more
clever and creditable is it in any one to point out a
feature of loveliness previously concealed or overlooked.
Shakspere, above all men—the joy and solace of millions
in days gone by, and who will be the joy and solace of
millions yet unborn—is the last in the world who should
be subjected to pert and ungrateful testing for defects.
Look up, rather, when in his presence, as you do at
York Minster, and hope that, with much reverence, you
may some day be able to appreciate the full splendour.
Not that we are to regard Shakspere as a kind of demi-
god, and to think even his failings amiable. They go

with other men's, and with our own, such as were he
living, he could expose as readily, and with still greater
reason. It is a matter of satisfaction, after all, that they
exist. For while we love, and admire, and honour, no
matter who, that may be good and worthy, it is always
pleasant to feel that our love and honour are given to
good old-fashioned flesh and blood, to a being of substance like our own, animated, sometimes perhaps
disfigured, by the same passions, subject to the same
deficiencies and failings. This is one of the grand
secrets of the universality of men's fondness for Shakspere.
While teaching us, as some author says, that love is the
best of all things, "he reconciles us to our own defects."
With Shakspere we are always at our ease. His forehead
touches the sky; his voice gathers tone from the
immortals, but he still walks upon the earth and is one
of ourselves. Deducting the one little exception referred
to, Shakspere, in regard to his botany, though making
no pretention to be scientific, may always be trusted—
herein, perhaps, standing alone, at all events as compared
with all earlier and all contemporary literature, and with
the great mass of the poets of later ages. That several
of his plant and flower names are vague, and have given
rise to much conjecture and speculation, and that one or
two are probably undeterminable, may unhesitatingly be
conceded. Sometimes this comes of default of contemporary illustration; sometimes, in all likelihood, of the
carelessness of copyists—for it is to be remembered that
there is no such thing in existence as an original *manu-*

script of any portion of the Shaksperean writings, and
that before being printed they were exposed to every
species of corruption. But when we have the unquestionably original words, pure and unvitiated, we can
always read in faith, an assurance so much the more
agreeable because sometimes, at the first blush, there
may be disposition to demur. Take, for instance, the
pretty song in *Love's Labour's Lost:*—

> When daisies pied, and violets blue,
> And cuckoo-buds of yellow hue,
> And Lady-smocks, all silver white,
> Do paint the meadows with delight. *

Gather a Lady-smock as you tread the rising grass in
fragrant May, and, although in individuals the petals are
sometimes cream-colour, as a rule the flower viewed in
the hand is lilac—pale, but purely and indisputably
lilac. Where then is the silver whiteness? It is the
"meadows," remember, that are painted. When, as
often happens, the flower is so plentiful as to hide the
turf, and most particularly if the ground be aslope, and
the sun be shining from behind us, all is changed; the
flowers are lilac no longer; the meadow is literally silver-
white. So it is always—Shakspere's epithets are like
prisms; let them tremble in the sunshine, and we
discover that it is he who knows best.

The actual number of different species of trees, plants,

---

* The above is the arrangement of the lines in the early editions;
wrongfully, it would seem, nevertheless, and therefore, in most
modern ones, corrected so as to give *alternate* rhymes, as in all the
other verses of the song.

and flowers introduced or referred to by Shakspere, is by
no means large, considering how many more must have
been familiar to him.   The reason we have already
adduced; he never went out of his way for an illustration,
or sought to advertise his knowledge by importing anything in the least degree uncalled for.   He must have
looked a thousand times at the scarlet corn-poppies, at
the forget-me-nots, and the water-lilies on the stream.
But they were not wanted, and he left them to other
people.   About a dozen wild or field flowers, an equal
number of garden flowers, a score of the plants usually
called "weeds," and three or four of our indigenous trees,
absorb nearly all the passages distinguished for their
poetic beauty; and not uncommonly, as in the scant
passages in ancient Greek dramatic verse where flowers
are mentioned, the names and pictures come in clusters.
The plays richest in botanical allusion are, as above-
mentioned, the *Midsummer Night's Dream*, the *Winter's
Tale*, *The Tempest*, *Love's Labour's Lost*, *Hamlet*, *Romeo
and Juliet*, and *King Lear*.   These, it will be perceived,
cover nearly the whole period of Shakspere's productive-
ness, so that we cannot suppose him to have been
inclined to such allusions more particularly at any one
time, sooner or later, than another.   Few or many, it is
not so much after all, what plants does he name, as after
what fashion does he name them?   Had he omitted the
cowslip and harebell, and given us instead the poppy
and the forget-me-not, it would have been just the same,
since to the master everything comes right.   The way or

nature of the mention is in itself very interesting, being so various. Sometimes they are cited as objects of simple beauty, serving to decorate the landscape, to throw sweetness on field and woodland, as in "When daisies pied." Sometimes they become capital descriptive adjectives, as in "her lily hand." Still more frequently they supply delightful comparisons,—

> Kate, like the hazel twig,
> Is straight and slender, and as brown in hue
> As hazel-nuts, and sweeter than the kernels.
> *Taming of the Shrew*, ii., 1.

Often, again, they serve the purpose, incomparably, of emblems;—

> In such a night,
> Stood Dido, with a willow in her hand,
> Upon the wild sea-banks, and waft her love
> To come again to Carthage.—*Merchant of Venice*, v., 1.

In some cases, yet further, they appear by elegant implication, without actual mention of the name, as in the "enchanted herbs" of the wily little Jewess in the same play; and in the beautiful picture in the *Tempest* of the "fairy rings" produced by the centrifugal spread of certain fungi;—

> You demi-puppets, that
> By moonshine do the greensome ringlets make,
> Whereof the ewe not bites.—(v., 1.)

To the classes indicated may be referred forty or fifty of the total number of plants referred to, and these, from the mode of handling, may be regarded as the eminently poetical ones. Nearly a hundred others are alluded to

in an ordinary, conventional, or prosaic way, sometimes
the fruit or other product being alone spoken of. Of
edible vegetables about thirty are mentioned; of fruits,
native and foreign, the same number; of spices and
medicines about twenty. Occasionally the references to
the plants of this second and much larger class are
fraught with some kind of curious lateral interest, either
historical, or identified, in deference to the popular folk-
lore of the time, with some quaint superstition; for the
wonderful tales told by Pliny, and diffused during the
dark ages, during which he was the great authority, were
still implicitly believed in, at all events by the crowd.
The seed of the shield-fern sprinkled over one's body, in
Shakspere, as in his predecessors, confers invisibility:—

> We have the receipt for fern-seed,
> We walk invisible.—*1st Henry the Fourth*, ii., 1.

The bay-trees die when calamity is at hand:—

> 'Tis thought the king is dead. We will not stay,
> The bay-trees in our country all are withered.
> *King Richard the Second*, ii., 4.

The mandrake, source of Rachel's famous dudhā'īm,
when torn out of the earth, shrieks so terribly—

> That living mortals hearing it, go mad.
> *Romeo and Juliet*, iv., 3.

We must be careful at all times not to assign or attach to
Shakspere anything which is not really his own, especially
when he simply echoes some popular story. These
quaint old fancies and superstitions rank with our own
current use of the myths and legends of antiquity.

The singers talk of Apollo, the dancers of Terpsichore, the tragedians of Melpomene. Ulysses' lotus-tree has never ceased to ripen its golden berries for our own personal suasion. Helen still awaits us with her cup of nepenthe, and bestows it with grace as kindly as when offered to weary Telemachus. The tales and superstitions in question are no more a part of Shakspere's veritable and personal botany, than the old Greek fables are a part of our own individual and private religion, therefore not to be mistaken for imperfect scientific knowledge, much less for credulity or folly. The principle is an important one. Not alone in connection with the plants of the ancient fabulists, but at many other times, Shakspere must be understood as speaking less in *propriâ personâ*, than in conformity with the tastes and habits of the age. To this circumstance is to be ascribed whatever in his writings is impure. When he wrote to please himself alone, it was always wisely and loveably; that which now offends was written, if not distinctly to please the public, at all events in full consistency with the spirit and the manners of the age, which permitted many things now distasteful; though it may still be asked perhaps, if nineteenth century ideas of refinement and modesty are not, in comparison with Elizabethan ones, in many cases shadowy and fictitious.

It is important to remember also that of many of the hundred or more of the second class of plants spoken of, Shakspere had no personal knowledge, citing them either from books or from hearsay. Like every one else he talks of things he had never seen. Shakspere never saw either a palm-tree or a cedar, and probably knew as little of the myrtle and the olive. In contemplating his botany all such plants as these have to be differentiated from the trees and flowers he knew as companions and loved so dearly, and should on no account be catalogued with them.

Chapter Second.

IN THE WOODS.

Love thyself last: cherish those hearts that hate thee;
Corruption wins not more than honesty—
Still in thy right hand carry gentle peace
To silence envious tongues. Be just and fear not.
Let all the ends thou aims't at be thy country's,
Thy God's, and truth's.—*King Henry the Eighth*, iii., 2.

IKE every true lover of nature, Shakspere is always at home in the Woods :—of these, as an artist, he never tires ; in the woods, as a skilful dramatist, he lays some of his most admired and poetic scenes. Shakspere's acquaintance with sylvan scenery was certainly much more intimate than with mountains, waterfalls, and other grand elements of inanimate nature. There is no reason to believe that he ever visited North Wales or the Lake District ; and the seashore must of necessity have

been unfamiliar, though he knew enough of it to give us
the immortal picture in *King Lear* :—

> The murmuring surge,
> That on the unnumber'd idle pebbles chafes,
> Cannot be heard so high;

that other, in *2nd King Henry the Fourth*,

> Yea, this man's brow, like to a title-leaf
> Betrays the nature of a tragic volume :
> So looks the strand whereon the imperious flood
> Hath left a witness'd usurpation;

and that lovely one in the *Tempest*, where with himself
we see the little children giving the wave their old-
accustomed summer challenge;—

> Ye elves of hills, brooks, standing lakes, and groves;
> And ye, that on the sands with printless foot
> Do chase the ebbing Neptune, and do fly him,
> When he comes back.

Spending his youth in the ancient and glorious forest-
shades of Warwickshire, and returning to them after his
London life, no wonder that trees hold a place so
distinguished in his imagery. It was under the boughs
of immemorial forest-monarchs that his imagination
found earliest nurture; and no pleasure that we can
conceive of as concurrent with his declining years can
have exceeded the calm delight with which he trod the
shaded pathways wherein he had gathered his first
impressions of the beauty of nature, and tasted the deep
joy of meditation. Not only were grand old trees a
daily spectacle during his boyhood: he was much *alone*
with them, as with most other elements of wild nature,

and thus peculiarly open to their influence. It is fortunate for us that he was so circumstanced. Mr. Ruskin somewhere remarks that the quietude of Shakspere's early intercourse with nature contributed in no slight measure to the perfection of mental power disclosed so marvellously at a riper age, and which the word "Shaksperean" is sufficient to denote. His walks were in scented meadows, where he would hear no voices but those of the birds, and by the smooth and lilied river, from which he would change to the green recesses of the forest. No other scenes were at his command, save in the village, and even here the prevailing condition would be one of tranquillity. But we must not think of Shakspere's forests from the woodlands of to-day. Except in Sherwood, and the older parts of William Rufus' famous plantation, we have little of the kind which in the Elizabethan times still existed in plenty. Wheat now grows upon many a broad acre which, when Shakspere wrote, was covered with timber, and not simply timber such as pleases a modern dealer, but magnificent and aboriginal forest, the like of which in England can never be seen again. Many of the trees now so common in England that they seem indigenous, the birch for example, and the Lombardy poplar, had not been introduced; and even the sycamore and the Norway spruce were known only in private pleasure-grounds. Shakspere's forest consisted of trees such as had given shelter to Caractacus, and the great mass of them would be majestic. Those which now occupy the place of the

c

aged Titans of the olden time, are in comparison small and juvenile. We may learn what the former were from the huge slices preserved in the best of the old Elizabethan mansions, Haddon Hall for instance. Happily, too, a few survive to tell their own story, dotted over the country, as the Cowthorpe oak, the Marton oak, Sir Philip Sidney's, and others of the well-named "memorial trees" of our island. Trees such as these must be thought of when we would understand in what kind of school Shakspere learned his forest-lessons. They were not received from saplings of only a century or two of birthdays, but from patriarchs.

Of the many beautiful scenes laid by Shakspere in the quiet of great woods and forests, the most charming are those in the *Midsummer Night's Dream*, and in *As You Like It*, that delicious pastoral, in which it has been said so truly, "he teaches us how to forget the painful lessons of life, in the contemplation of faithfulness, generosity, and affection." The chief part of the action in each of these matchless pieces lies amid trees; and it is worth noting that it is in these two that Shakspere most wins upon the heart that delights in peace. Nowhere are we nourished more exquisitely by his humane and dulcet wisdom than when listening to him among the trees which bore "love songs on their barks" :—

> Now, my co-mates and brothers in exile,
> Hath not old custom made this life more sweet
> Than that of painted pomp? Are not these woods
> More free from peril than the envious court?
> Here feel we but the penalty of Adam,

> The seasons' difference; as the icy fang,
> And churlish chiding of the winter's wind,
> Which, when it bites and blows upon upon my body,
> Even till I shrink with cold, I smile, and say
> This is no flattery; these are counsellors
> That feelingly persuade me what I am.
> Sweet are the uses of adversity,
> Which, like the toad, ugly and venomous,
> Wears yet a precious jewel in his head.
> And this our life, exempt from public haunt,
> Finds tongues in trees, books in the running brooks,
> Sermons in stones, and good in everything.

Nowhere, either, are we touched more tenderly with thought of what is gracious and chivalric than when with Helena in "another part of the wood"—not now Arden, but that lovely one "near Athens."

> If you were civil, and knew courtesy,
> You would not do me thus much injury.
> Can you not hate me, as I know you do,
> But you must join in souls to mock me too?
> If you were men, as men you are in show,
> You would not use a gentle lady so.
>
> .   .   .   .
>
> A trim exploit, a manly enterprise,
> To conjure tears up in a poor maid's eyes!
> > *Midsummer Night's Dream*, iii., 2.

Nature, says a great essayist, is "coloured by the spirit." Hence, in the former play, to Orlando, who carries sorrowful thoughts with him, all for a time is rude, waste, and disheartening:—

> Speak you so gently? Pardon me, I pray you,
> I thought that all things had been savage here,
> And therefore put I on the countenance
> Of stern commandment. But whate'er you are,

That in this desert inaccessible,
Under the shade of melancholy boughs,
Lose and neglect the creeping hours of time;
If ever you have look'd on better days,
If ever been where bells have knoll'd to church,
If ever sat at any good man's feast,
If ever from your eyelids wiped a tear,
And know what 'tis to pity, and be pitied,
Let gentleness my strong enforcement be.
                                        As You Like It, ii., 7.

The "colour" laid on by sadness, Emerson might have
gone on to tell us, is rarely other than one that presently
fades and disappears, seeing that the grand function of
nature is to refresh and revive the heart.   Always ready
with an echo for joyfulness, she refuses to sustain the
mournful.   "In the woods," according to his own
experience, "we return to reason and faith.   There I
feel that nothing can befall me in life—no disease, no
calamity, which (leaving me my eyes) nature cannot
repair."   Old Homer represents Achilles as regaining his
lost composure through playing on his harp to the sound
of the sea.   Shakspere keeps us to the forest; he knew
where best to lay his encouraging scene.   He shows
withal his consummate art in interweaving with wild
nature the still more potent realities of human emotion.
Orlando soon discovers that the woods, after all, are *not*
melancholy, and henceforward we ourselves enjoy them
threefold.   Where Rosalind breathes, how can any place
be sad—Rosalind, darling maid, one of the quintette of
Shaksperean women to be compared to any one of whom
is compliment enough for any of her sex that ever lived—

Rosalind, who, in her boy's clothes, makes believe that she does not know who writes the verses, or for whom they are intended. Women love nothing better than to be able to feign ignorance of the emotions and actions they hold most dear. Shakspere, in this charming episode, shows once again that the poet rightfully so named, is, as the ancients said, neither man alone, nor woman alone, but *homo*.

Shakspere did not care to learn much about what a botanist would call the "species" of trees. It is doubtful if he knew familiarly more than half-a-dozen different kinds, including even the smaller ones of the hedgerow. He never once mentions the beech or the abele, and even the ash and the elm hold no place in his landscape pictures. But how quick and accurate his perception of the phenomena of their life, and of the part they play in the universal poësy! This is the kind of knowledge to be most envied, for it is that to which comparison of forms and colours never reaches. A single instance of the first will suffice. We are all well acquainted with the great outpour of foliage which marks the spring. Are we as apt to notice that in July there is in many kinds of trees a lull or pause, followed in a little while by a distinctly second series of twigs? Sometimes they are yellowish, sometimes roseate, occasionally of a warm and shining ruddy hue, looking like bouquets among the green.

And never, since the *middle summer's spring*,
Met we on hill, in dale, forest, or mead,

By pavèd fountain, or by rushy brook,
Or on the beachèd margent of the sea,
To dance our ringlets to the whistling wind.
Midsummer Night's Dream, ii., 2.

The trees which in our own country are most especially
apt to produce these beautiful midsummer shoots, are
the oak and the sycamore, the reason being found in the
exceptionally large number of lateral buds which these
two are prone to develop, in a circlet, around the
terminal bud, one of any such circlet, wherever met with,
always anticipating its neighbours.   Shakspere could
scarcely have been able to observe them in the sycamore,
as in the time of Elizabeth this tree was only beginning
to be known.   When he wrote these charming lines his
thoughts were with Old England's oak.

Turn now to the poësy proper.   In the sunshine of
high summer,

The green leaves quiver with the cooling wind,
And make a chequer'd shadow on the ground.
Titus Andronicus, ii., 3.

Then we are asked to note how quiet they can be :—

The moon shines bright ; in such a night as this,
When the sweet wind did gently kiss the trees,
And they did make no noise —*Merchant of Venice*, v., 1.

Presently the breeze quickens:—

The southern wind
Doth play the trumpet to his purposes,
And by his hollow whistling in the leaves
Foretells a tempest, and a blustering day.
1st Henry the Fourth, v., 1.

Autumn approaches now :—

> I have lived long enough : my way of life
> Is fall'n into the sear, the yellow leaf.—*Macbeth*, v., 3.

Lastly, mark the observation, so consummately accurate, of the fact not more true in botany than so admirably employed as an image, that a tree never casts its principal or larger leaves, till decay of everything is imminent :—

> When clouds are seen, wise men put on their cloaks ;
> When *great leaves fall*, the winter is at hand ;
> When the sun sets, who doth not look for night?
> > *King Richard the Third*, ii., 3.

THE OAK.

Every country has its "forest monarch." In England, this proud title is rightfully accorded to the oak—the majestic *Quercus Robur* which in associations as well as figure and attributes, owns no rival. Many circumstances contribute to the supremacy. The dimensions, when full-grown, exceed those of every other British tree. The outline or profile, though in general character so determinate that to mistake an oak is impossible, is inexhaustibly various. The trunk, huge and massive, though never aspiring, holds the place among foresters which the Norman pillar does in cathedral architecture. The lower boughs, spreading horizontally, often so nearly touch the ground as to allow of our gathering not

only the acorns, but their pretty tesselated cups. The
rugged bark is peculiarly open to the embroidery of
delicate mosses, green and golden; and just high enough
to be secure from touch, there is often a beautiful tuft
of spangled polypody, in winter a cheerful ornament
unknown to any other English tree. Neither is there in
England any tree that presents so wonderful a diversity
of leaf-outline, or a richer variety of summer tint. The
autumnal hue is scarcely exceeded even by the beech
and the elm; and when these beautiful leaves, their
tasks completed, pass away, not afraid of dying, the
birch itself does not disclose secrets of loveliness more
delectable. It should never be forgotten in regard to
deciduous trees in general, and in reference to the oak
most particularly, that however delightful the spectacle at
midsummer, when clothed with foliage, slaking its mighty
thirst in the well-pleased sunshine, the inmost form is
learned only in winter, or when we are reminded of the
classic fable of Mount Ida.

To Shakspere, without question, all these features, the
pretty minor ones as well as the noble, must have been
familiar. His imagination must needs also have been
influenced by the noble ones, an almost daily spectacle,
and to a degree it is now very difficult to estimate; and
if the minor ones receive no express or exact mention, it
is simply because the verse was complete without. How
beautiful the picture of the aged tree by the water side
in peaceful Arden, the sturdy roots laid bare by the
washing away of the earth that once protected them:—

> He lay along,
> Under an oak whose antique root peeps out
> Upon the brook that brawls along this wood.
>> *As You Like It*, ii., 1.

That one again, in the same play, of the mighty tree,

> Whose boughs were moss'd with age,
> And high top bald with dry antiquity.—(iv., 3.)

If we are so constituted as to delight in trees while they are youthful, and again, while in the full vigour of existence, what is there for which we can be more grateful than the capacity to enjoy them when old and grey? Mr. F. Heath points out very happily that it has pleased God so to adjust the lives of trees to the lives of men, that no generation of mankind comes on the ground without finding both the promise and the perfection of the divine handiwork in the vegetable world side by side. Man has all the stages of tree-life always before him:—Shakspere takes care that this at least we shall not forget.

The moss and the antiquity re-appear in *Timon of Athens*:—

> Will these moss'd trees,
> That have outlived the eagle, page thy heels,
> And skip when thou point'st out ?—(iv., 3.)

That a very considerable protion of *Timon of Athens* was not written by Shakspere is generally admitted. *Timon*, like *Pericles* and *Titus Andronicus*, came fundamentally from another pen, Shakspere altering and amending. To decide exactly which parts belong to the inferior authors is work for critics who may be competent. All that can

be done while we are "in the woods" is to take the play
as it stands, and for what it is, bearing in mind that
Shakspere in his boundless knowledge of the mingled
web of human nature, allows even the most ignoble at
times to utter truths;—thus, that although Apemantus,
who speaks, is a character when put in comparison with
the generous Timon, mean and heartless—it was probably
Shakspere himself who assigned to him the beautiful
adjuration :—

> Shame not these woods,
> By putting on the cunning of a carper.—(iv., 3.)

Here, in the woods, he means, as we have already
learned from the most illustrious of Dukes, regrets and
lamentations are out of place :—remember that in the
woods we are upon consecrated ground; it is in the
woods that the heart finds solace and repair; shame
them not with murmurs.

Shakspere refers frequently to the prodigious strength
and solidity of the oak :—

> I have seen tempests when the scolding winds
> Have rived the knotty oaks.—*Julius Cæsar*, i., 3.
>   . Merciful heavens !
> Thou rather, with thy sharp and sulphurous bolt,
> Split'st the unwedgeable and gnarlèd oak,
> Than the soft myrtle.—*Measure for Measure*, ii., 2.
>   To the dread rattling thunder
> Have I giv'n fire, and rifted Jove's stout oak
> With his own bolt.—*Tempest*, v., i.

"Jove's," because with the ancients, the oak was dedicated to Jupiter, as the myrtle was to Venus, the laurel
to Apollo.

The same qualities recommend it for use in simile, and for metaphor, as in *Much Ado*, "An oak with but one green leaf upon it would have answered her." (ii., 1.) So in *Coriolanus:*—

> He that depends
> Upon your favours, swims with fins of lead,
> And hews down oaks with rushes.—(i., 1.)

And in *Love's Labour's Lost*, iv., 2.,—lines which occur also in the *Passionate Pilgrim:*—

> Though to myself forsworn, to thee I'll faithful prove,
> Those thoughts to me were oaks, to thee like osiers bowed.

Pauline, in the *Winter's Tale*, ii., 3.; Montano, in *Othello*, ii, 1.; Arviragus in *Cymbeline*, iv., 2.; and Lear, iii., 2., turn to it for the same species of illustration. There are similar examples in *Troilus and Cressida*, i., 3.; and others in *Coriolanus*, v., 2 ; v., 3.; while in *3rd Henry the Sixth*, ii., 1., comes that old familiar and beautiful picture of the final reward of perseverance:—

> And many strokes, though with a little axe,
> Hew down, and fell the hardest-timbered oak.

Individual trees mark the scene of assignations and adventures, as in the *Midsummer Night's Dream*, "At the duke's oak we meet"; and in the mirth-provoking *Merry Wives*, that joyous play, the finest example ever produced of the purely and thoroughly English local drama, in which we are introduced to the famous legend of Herne the Hunter:—

> There is an old tale goes that Herne the Hunter,
> Some time a keeper here in Windsor forest,
> Doth all the winter-time, in still midnight,
> Walk round about an oak, with great ragg'd horns.—(iv., 4.)

There is no need to quote the remaining passages, seven or eight in all. Suffice it to say that this celebrated tree—the individual, at all events, always pointed to as "Herne's Oak," existed till as late as 1834, and continued even till then, to bear acorns just as in the time of "Sweet Anne Page." It then succumbed, partly through exhaustion, partly to the effects of tempest, though the dead ruins remained in their place up to 1863.

The Romans gave chaplets of oak to distinguished soldiers. Reserved for the most valiant, and as the reward of special acts of heroism, they were the antetype of our own "Victoria Cross," and are fittingly mentioned in *Coriolanus:*—

> He comes the third time home with the oaken garland.—
>
> (ii., 1.)
>
> He prov'd best man i' the field, and for his meed,
> Was brow-bound with the oak.—(ii., 2.)

In one curious instance "oak" appears to be a copyist's error for "hawk":—

> She that so young could give out such a seeming
> To seel her father's eyes up close as oak.—*Othello*, iii., 3.

When falconry was a royal sport, newly-captured hawks had their eyelids stitched together, so as to accustom them to the hood. The operation, called "seeling," supplies a grand metaphor in *Macbeth*:—

> Come, seeling night,
> Scarf up the tender eye of pitiful day,

and the same, in all probability, is referred to in this otherwise obscure *Othello* passage.

Acorns, the fruit of the oak, are mentioned upon some half-dozen occasions. "I found him," says Celia, "under a tree, like a dropped acorn." Rosalind is ready for her: "It may well be called Jove's tree, when it drops forth such fruit." Then comes the picture drawn from chivalry or the wars: "There lay he, stretched along like a wounded knight," followed up by Rosalind's sweet sympathy, as earnest as her love, "Though it be pity to see such a sight, it well becomes the ground," a most beautiful phrase, to be interpreted as "embellishing" it; just as at the end of *Hamlet*, when the dead are lying so thick upon the stage:—

> Such a sight as this
> Becomes the field, but here shows much amiss.

It is easily understood if we think of the similar use of the word in our own current vernacular, as when it is said so fittingly that a blush becomes a woman.

In the *Midsummer Night's Dream*, the cups give shelter to the fairies:—

> All their elves for fear,
> Creep into acorn-cups, and hide them there.—(i., 1.)

The same, in the *Tempest*, are to serve as food for Ferdinand, whose deserts, for the time being, are only those of the Prodigal Son:—

> Sea-water shalt thou drink : thy food shall be
> The fresh-brook mussels, withered roots, and husks,
> Wherein the acorn cradled. —(i., 2.)

Allusions of similar character in the *Tempest*, i., 2, *Timon of Athens*, iv., 3., and a few others of quite subordinate

importance, being added to the above, the total number
of references to the oak by Shakspere, appears to be
thirty-one, or excluding the repetitions in the *Merry
Wives*, twenty-four.   No other tree is mentioned so
often, and thus, upon his own showing, it was his
favourite; though we must not forget that the oak has in
all ages held a front place in metaphor, the various
names under which it appears denoting several species
not British, as in the case of the Hebrew *'allōn*, and the
Greek δρῦς, the word employed in that famous line in the
Odyssey:—

"For thou art not of the oak of ancient story."

Chapter Third.

THE WILLOW.

This above all, To thine own self be true,
And it must follow, as the night the day,
Thou can'st not then be false to any man.

Hamlet, i., 3.

HE willow, like the oak, is placed before us by Shakspere both as an object of natural beauty and as an emblem. It is interesting to observe, at the outset, that, excepting a slight reference in Virgil to the form and colour, he is the first poet by whom it is substantially so employed.

QUEEN : Your sister's drowned, Laertes !
LAERTES : Drowned ! O, where ?
QUEEN : There is a willow grows ascaunt the brook,
That shows his hoar leaves in the glassy stream.
There, with fantastic garlands did she come,
Of crow-flowers, nettles, daisies, and long-purples,
That liberal shepherds give a grosser name,
But our cold maids do dead men's fingers call them.

There, on the pendent boughs, her coronet weeds
Clambering to hang—an envious sliver broke ;
When down her weedy trophies and herself
Fell in the weeping brook.   Her clothes spread wide,
And mermaid-like, awhile they bore her up,
Which time she chanted snatches of old tunes,
As one incapable of her own distress,
Or like a creature native and indued
Unto that element.   But long it could not be,
Till that her garments, heavy with their drink,
Pull'd the poor wretch from her melodious lay,
To muddy death.—*Hamlet,* iv., 7.

The tree particularly alluded to in this most beautiful
and tender description, is that well-known lovely ornament of the river side—the white willow, *Salix alba*—a
species similar to the common *S. fragilis,* or "Crack
willow," but at once distinguished by the long and
narrow leaves being overspread with shining silvery hairs.
Enamoured of quiet streams such as Shakspere was
familiar with in Warwickshire, upon the Avon it still
accentuates many a reach, and to-day we may see it
reflected just as *he* did.   When growing on the very
margin, and at a point where the current newly presses,
the trunk cannot help but lose its hold upon the soil,
which is undermined and worn away, so that at last it
quite leans over, the boughs then often forming a light
canopy, beneath which the little fishes play.   Favourably
circumstanced, it attains a stature of thirty feet, and the
branches are then strong enough to bear the weight of
one who climbs them.   Mark now the supreme art of
the master in telling us that the water was deep enough

Drawn & Etched by Tho. Letherbrow.

The Shrine

for the drowning of the poor soul, without saying so in express terms. Shallow streams, such as cannot drown, are always more or less rippled. They cannot possibly serve as mirrors; they do not reflect even the stars; it is only for the deep and tranquil to be "glassy," and to let objects such as the willow reappear.

There are other pictures to be contemplated. The foliage of this beautiful tree, though like the silvery head of a Nestor, it gleams in the ray of the sun, is wanting in the peculiar sheen of the oleaster. Hence the epithet "hoar," subdued or greyish-white, already immortalized by Chaucer:—

> Though I be hoar, I fare as doth a tree
> That blosmeth ere the fruit y-woxen be ;
> The blosmy tree is neither drie nor ded ;
> I feel me nowhere hoar but on my hed ;
> Mine harte and all my limmès ben as green
> As laurel through the year is for to seen.

Note also, "mermaid-like." "Mermaid" in Shakspere's time was synonymous with "siren":—

> O train me not, sweet mermaid, with thy note,
> To drown me in thy sister's flood of tears,
> Sing, siren, for thyself.—*Comedy of Errors*, iii., 2.

So in the splendid description of Cleopatra's barge:—

> The barge she sat in, like a burnished throne,
> Burn'd on the water : the poop was beaten gold ;
> Purple the sails, and so perfumèd, that
> The winds were love-sick with them.

> Her gentlewomen, like the Nereides,
> So many mermaids, tended her i' the eyes,
> And made their bends adornings.
> *Antony and Cleopatra*, ii., 2.

D

So in *Venus and Adonis,*

> Thy mermaid's voice hath done me double wrong.

So, too, when Oberon relates how once he heard a mermaid,

> On a dolphin's back,
> Uttering such dulcet and harmonious breath,
> That the rude sea grew civil at her song.

This last passage supplies clear proof of the identity of meaning, since the mermaid of fable is in its lower half like a fish, therefore incapable of taking such a position. The sirens, on the other hand, held the complete human form.

The death of Ophelia (represented in pictorial art by Mr. Millais, in a work not less well known than truthful to all that the highest poetry and the imagination can desire,) has the further and most profoundly tender interest to the student of Shakspere, which comes of its being the only instance in the whole of his writings of life lost by drowning. In all, he has about ninety deaths, many of them violent and shocking. It was fitting that one whose existence had been pure as a snow-flake, and as easily dissolved, should pass away when her mind was already gone, thus calmly and silently, and by means so gentle. No struggle, no fear, not even consciousness of what presently must · needs happen, "incapable of her own distress,"* when her eyes close it is like the shut at sundown of the water-lilies.

---

* Incapable = unconscious. " Distress," as when seamen talk of "stress of weather."

"Ascaunt," it may be proper to say, is the reading in the second, third, and fourth quarto editions of *Hamlet*, 1604, 1605, and 1611, also in the fifth, undated, though in the first folio, 1623, changed to "aslant." The two words have precisely the same meaning, and which of them shall be adopted is purely a matter of taste.

There is no other allusion in Shakspere to the willow as an object of the landscape, not even to the common *fragilis;* though we find several little references to the smaller kinds of Salix known as osiers and withies, forms of the *viminalis*, and which are still used by basket-makers, as in the olden time for shields and coracles. Friar Lawrence's "willow-cage," in *Romeo and Juliet:*—

> I must fill up this willow-cage of ours
> With baleful weeds and precious juicèd flowers—

is to be understood as a little calathus, such as Persephone carried. Cassio's "twiggen bottle," in *Othello*, ii., 3, was one enveloped in wicker-work, something after the fashion of a Florence flask. When Viola, talking to Olivia, in *Twelfth Night*, i., 5, says Ah! did it fall to *my* lot to show devotedness, I would

> Make me a willow-cabin at your gate,

she means that rather than be away from her love, she would be content to live at his door with shelter no better than that of a bird-cage. In *Love's Labour's Lost*, the withy, being so pliable, is an emblem of weakness of character:—

> Those thoughts to me were oaks, to thee like osiers bowed.

In *As You Like It*, iv., 2, the natural habitat of the plant is mentioned:—

> West of this place, down in the neighbouring bottom,
> The rank of osiers by the murmuring stream
> Left on your right hand, brings you to the place.

So again in the *Passionate Pilgrim*:—

> When Cytherea all in love forlorn,
> A longing tarriance for Adonis made
> Under an osier growing by a brook.

In this last, however, "osier" is probably used in the sense of willow.

When introduced metaphorically, the willow appears as in the *Merchant of Venice*, v., 1:—

> In such a night,
> Stood Dido with a willow in her hand
> Upon the wild sea-banks, and waft her love
> To come again to Carthage.

Nay. The original tale represents her as too busy upon her watch-tower, "beating her fair bosom with repeated blows, and tearing her golden locks."

> Terque quaterque manus pectus percussa decorum,
> Florentesque abscissa comas.

Shakspere is right all the same. Whether she beat her bosom or not, he simply employs in these beautiful lines, in another form, the celebrated figure in the Psalms— "By the rivers of Babylon there we sat down; yea, we wept when we remembered Zion; we hanged our harps upon the willows in the midst thereof," in the "midst" meaning in the middle of the city. That no literal fact is intended by the Psalmist, any more than in the Dido story, scarcely needs the saying. The captive Hebrews no more suspended their harps upon willows or any other trees, than the forests of Palestine "clapped their hands," or than the valleys, covered with corn, "shouted for joy." The phrase is purely figurative, harmonizing both in spirit and form, with a thousand others in the inspired volume. To all appearance, "hanging up the harps" was with the Hebrews a current saying, adopted probably from Job xxx., 31, "My harp is turned to mourning, and my organ"—some kind of musical pipe—"into the voice of them that weep." Happy the day when in the presence of all such passages, secular as well as sacred, people shall learn how to discriminate, before they accept, between the "letter which killeth, and the spirit which giveth life."

To get at the sense of the Shaksperean words, we have thus to ascertain, first, the sense of those contained in the psalm, and this demands as the earliest step, the determination of what is really meant by "willow." The current idea is that both Shakspere's willow and the scriptural one are the common "weeping willow," *Salix Babylonica*. But this is a native of China; it was not carried westward until a comparatively recent period, and received its specific Latin name in error. The Hebrews knew no more of the *Salix Babylonica* than Shakspere did, and to him it was quite a stranger, not arriving in this country until many years after his death. The original Hebrew word (usually transliterated *'arabhim*) was, like many other Old Testament Hebrew botanical terms, one of collective signification. It included the *Populus Euphratica;* probably that charming shrub, the silver elæagnus; probably also the agnus-castus and the oleander, and not impossibly, a genuine Salix. But it certainly did not mean willows only. The willow, strictly so called, had moreover, with the Hebrews, a distinct appellation of its own, *tsaphtsāphāh*, the word employed in Ezekiel xvii., 5. The *'arabhim* are mentioned in the Hebrew upon five different occasions. In two instances the associations are distinctly joyous.* In one there is nothing sorrowful;† in the fourth the word is simply part of a geographical name;‡ the fifth, the verse in the psalm, is without any legitimate reason, by the Septuagint translated ἰτέαις, the word used by old Homer for osiers or withies, and which also have their own proper Hebrew name, *yĕtharim*, as in Judges xvi., 7. The translators of the Vulgate followed suit, and put *salices* in all the passages. Wiclif, translating from the Vulgate, has in his earlier version, "In withies in the myddes of it, wee heengen vp oure instrumens;" and in the later one, "In salewis in the myddil thereof, we hangiden vp oure organs." Coverdale (1536), and Tyndall (1549), return to the indefinite expression "upon the trees." "Willows" appears first in the Geneva edition, 1560 (the first in which the chapters were divided into verses), and as this was the

---

* Levit., xxiii., 40.   Isaiah, xliv., 4.   †Job, xl., 22.
‡ Isaiah, xv., 7.

edition chiefly used in private houses and families during the reign of Elizabeth, and up to 1611, in all likelihood it was the form of the Bible in which Shakspere was so well read. It is the Geneva, at all events, which we are to regard as the parent-source, after the Septuagint, of the association now under review. Nature, as we have seen, is "coloured by the spirit." Haunting the calm and silent pathway by the stream, which in invitation to the unhappy is so like the "shadowy desert:"—more or less inclined to droop or "weep;" having leaves so long, narrow, and acutely pointed, that when the rain falls on them, trickling gently, it drips from the extremities like tears—no wonder that these elegant trees, the *'arabhim*, in their various kinds, seemed to the sad-hearted Hebrews representative of their own feelings. To describe unhappiness under the exquisite figure of hanging their harps upon the branches, was but another little step. Orientals could hardly do otherwise. At last, almost imperceptibly, the phrase transmutes, as we have just seen, into hanging them upon "willows." Language, always ready to utilize and give permanence to a pretty image, thenceforwards makes the willow the symbol of grief in general, and especially of the grief of disappointed love; and so at last we have the willow which Dido waft in vain, cut from no tree in nature, yet quite as true and real to the imagination. No allusion to the willow as an emblem of sorrow occurs in any one of the Greek or Latin poets, showing that the origin of the usage is supplied purely by the Septuagint. Observe

also, before we leave her, that the tale of unhappy
Dido is accepted by Shakspere as told by Virgil, which
is untruthfully.   Dido, in reality, never saw Æneas.
She lived two hundred years before the time of the
famous Trojan, and not upon her was it that he bestowed
either his caresses or Helen's robe.   Never mind.   Virgil
must answer for himself, though one regrets that he
should have so wantonly sullied the memory of a noble
lady: whatever the historical facts, the poetry remains
intact, and it is in this for us to rejoice.   It was no new
event either for even Virgil to describe.   The pictures in
the great poets are contemporaneous with all ages.   If
one "bonnie Doun" has been sung of, a thousand others
unrecorded have helped to give a summer-evening
paradise, to end after the same fashion as hers who
plucked the "rose."   As long as the world endures
Œnonë will point to her lettered poplar, and the willow
be waft anew upon "wild sea-banks."

We are never at liberty, under any circumstances, to
put anything into Shakspere of our own.   To take care,
at the same moment, that we miss nothing; to seek at
all times to elicit the charms of his undertones, is still
quite legitimate, perhaps our bounden duty.   The
Ophelia passage itself seems another utterance of the
above idea.   She might have died on land, and in any
other way, but the willow "ascaunt the brook" gives the
event a tenfold pathos to the imagination.

In *Othello* Shakspere gives us a varied rendering of
the celebrated old ballad printed in the Roxburghe

collection (i., 171), "The complaint of a lover forsaken by his love." He adapts it, quite legitimately, so as to suit the new character, that of Desdemona the ill-fated :—

> My mother had a maid called Barbara;
> She was in love; and he, she loved, proved mad,*
> And did forsake her: she had a song of willow,
> An old thing 'twas, but it express'd her fortune,
> And she died singing it. That song, to-night,
> Will not go from my mind; I have much to do
> But to go hang my head all at one side,
> And sing it like poor Barbara.

Then comes the song itself, with its sorrowful refrain, "willow, willow, willow." By and by, after other troubles, a line or two of it comes from the lips also of dying Emilia, sustaining the tender imagery to the end :—

> I will play the swan,
> And die in music, "Willow, willow, willow!"

Even when hope is gone, says Victor Hugo, and despair comes, Song remains.

One other serious reference to the willow in this connection occurs in Shakspere, *3rd King Henry the Sixth*, iii., 3, repeated in iv., 1 :—

> Tell him, in hope he'll prove a widower shortly,
> I'll wear a willow garland for his sake.

And lastly, we have it introduced jocularly, when in *Much Ado*, ii., 1, Benedick, laughing in his sleeve at Claudio, proposes to go to the "next" or nearest willow, for a garland, "as being forsaken." These are all the

---

* Wild, inconstant, heartless.

Shaksperean allusions, but it would be easy to cite others from contemporary authors, as when Spenser gives the willow to the "forlorne paramour," and when Montanus, in Lodge, at last about to wed Phœbe, throws away his "garland of willow." Chaucer had already used it in the construction of the funeral pyre in *Palamon and Arcite.*

THE YEW (*Taxus baccata*).

That Shakspere was well acquainted with the yew there can be no doubt. Though not so common as the oak or the beech, no part of our island is unpossessed of this famous tree. A genuine ancient Briton, it occurs in the secluded parts of all old forests. More than any other— excepting, perhaps, the mountain-ash—it is noted for its love of loneliness. Wild and desolate places are enjoyed; rooted in the crevices, it clings even to vertical rocks, where it can be touched by no other plant, and will have no companions but the sunshine and the rain. Compared with other British trees, it is the most massive and imperturbable of all, knowing nothing of storm or tempest, heat or cold. In early summer its perennial darkness is relieved awhile by pretty sprays of new-born verdure; and in late autumn the boughs are decked with scarlet berries, which the birds soon take away. At all other times the yew presents the ideas alone of antiquity and imperishableness, and if we do not exactly admire its ancient form, none can fail to recognize in it a grandeur almost unique.

The classical poets connected this tree with the mournful side of death. Silius Italicus, in his description of the nether world, places a yew-tree in the midst (xiii. 595-6). Claudian goes so far as to put torches made of the wood in the hands of the furies *(Rapt. Pros.* iii., 386). Christianity took the opposite view, finding in it the symbol of Immortality—the cheerful side of death—or that of which all Christians think primarily and with most energy. In the olden times every village church in our island would seem either to have been erected near a yew, or to have had one planted alongside. The countless extant examples in rural graveyards bear witness to the ancient practice, and to Shakspere the reverend custom must have been no less familiar than to ourselves. He knew the tree best, in all likelihood, as an inmate of such enclosures. Hence the introduction of the yew in *Romeo and Juliet:*—

> Under yon yew-trees lay thee all along,
> Holding thine ear close to the hollow ground,
> So shall no foot upon the churchyard tread
> (Being loose, infirm, with digging up of graves),
> But thou shalt hear it.—(v., 3.)

Balthasar, again, in a later part of the scene, says:—

> As I did sleep under this yew-tree here,
> I dreamt my master and another fought,
> And that my master slew him.

Shakspere constantly assigns to other countries and ages the customs and usages of his native land—of this, in due course, we shall have many illustrations—that he should place yew-trees in a graveyard at Mantua would thus be quite natural; but the practice above indicated would seem to have been very generally followed in Christian Europe during the Middle Ages, so that in the present place, there is, at all events, no inconsistency.

Identified in this beautiful manner with the Christian idea of futurity, it was natural, again, that sprigs of yew should be employed in funeral ceremonies:—

> My shroud of white, stuck all with yew,
> O ! prepare it !—*Twelfth Night*, ii., 5.

Immortelles and "everlastings" were not grown in the Elizabethan times. The first ever seen in England arrived only during the reign of Charles I. Few of the modern favourites have been known more than a century, if so long. Wreaths made of these glossy flowers waited for the rites of three hundred years later, or, instead of "sprigs of yew," perchance we might have had "amaranths." Contemplating the possibility, there is new reason to be glad that Shakspere lived just when he did, unoppressed by the lore of to-day, embosomed in the simplicities.

Though the juice of the sweet and viscid berries is not harmful, the seeds of the yew, and the leaves, are deadly poison. Taken in connection with the sombre appearance of the tree, and the profound shadow it casts, this, we may be sure, accounts sufficiently for the unhappy associations in which it is always found in ancient verse. In any case it explains the allusion in *Macbeth*, where the witches, in their gloomy cave, the cauldron boiling in the middle, prepare the mixture that is to give effect to their accursed sorcery:—

> Double, double toil and trouble,
> Fire, burn, and cauldron bubble.
>
> .   .   .   .
>
> Eye of newt, and toe of frog,
> Wool of bat, and tongue of dog,
> Adder's fork, and blind-worm's sting,
> Lizard's leg, and owlet's wing.
>
> .   .   .   .
>
> Liver of blaspheming Jew,
> Gall of goat, and slips of yew.—(iv., 1.)

The very specially interesting identification of the yew with the idea of venom comes to the front, however, in that fearful scene in *Hamlet* where the spirit of the murdered king describes the circumstances of his death. Note here that Shakspere's way of dealing with the great truth of the spiritual body of man forms one of the most striking adjuncts of his philosophy. Shakspere's belief is one in which the wise and good of all ages have concurred, and is set aside neither by ridicule nor incredulity. Men whose opinion is worth having never at any time deny even the simplest proposition till they are prepared with positive evidence to the contrary. "Hamlet," says the murdered monarch, addressing his son—

> 'Tis given out that sleeping in mine orchard,
> A serpent stung me; so the whole ear of Denmark,
> Is by a forgèd process of my death
> Rankly abused; but know, thou noble youth,
> The serpent that did sting thy father's life
> Now wears his crown!

> .   .   .   .
> Sleeping within mine orchard,
> My custom always of the afternoon,
> Upon my sècure hour thy uncle stole,
> With juice of cursed *hebenon* in a vial,
> And in the porches of mine ears did pour
> The leperous distilment.
>
> .   .   .   .
> Thus was I, sleeping, by a brother's hand,
> Of life, of crown, of queen, at once despatched.
>
> .   .   .   .
> No reckoning made, but sent to my account,
> With all my imperfections on my head.—(i., 5.)

What kind of liquid poison Shakspere intended by "hebenon" has been a subject of much conjecture, opinions oscillating chiefly between "henbane" and poisons in general, those who hold the latter view overlooking the minute description of the symptoms and pathological results. The word in question is a varied form, not of "henbane," or, as some suppose, of "ebony," but of the name by which the yew is known in at least five of the Gothic languages; the name which appears in Marlowe, Spenser, and other writers of the Elizabethan era as "hebon,"

> In few, the blood of Hydra Herne's bane,
> The juide of hebon, and Cocytus' breath,
> And all the poisons of the Stygian pool,
>
> *Jew of Malta*, iii., 4,

and which in the first quarto itself is spelt hebona. "The yew," says Lyte, translating Dodoens, "is called in high Dutch ibenbaum, and accordingly, in base Almaigne, ibenboom." "This tree," he goes on to say, "is

altogether venomous, and against man's nature. . . .
Such as do but only sleepe under the shadowe thereof
become sicke, and sometimes they die" *(Herbal,* 1578).
The extract is used, he says further on, "by ignorant
apothecaries, to the great perile and danger of the poor
diseased people" (p. 768). From the latter sentence we
may gather how the murderer was enabled to possess
himself of the deadly juice, which he is not to be supposed as preparing with his own hands, but as procuring
from one of the herb-doctors who kept it for sale. Why
Shakspere should say "hebenon" instead of "yew" does
not appear, nor does it signify. The scene being laid at
Elsinore, perhaps he was careful to employ a word
believed or known to be Danish.*

The ancient celebrity of the wood of the yew for
archers' bows needs no illustration. "Ityræos," says
Virgil, "taxi torquentur in arcus;"—"the yews are bent
into Ityræan bows" *(Georgic* ii., 448). Englishmen need
think only of Agincourt and Creçy. In *Richard the
Second* it is adverted to in a very curious passage, the
idea of the mortal certainty of the well-aimed arrow
being adjoined to that of the action of the poison:—

> The very beadsmen learn to bend their bows
> Of *double-fatal* yew against thy state. —(iii., 2.)

---

* The above teaching as to the true sense of "hebenon" has been
before my pupils for at least twenty-five years. It was with great
pleasure that I saw my views confirmed in the report of a paper
read by the Rev. W. A. Harrison, before the London Shakspere
Society, May 12th, 1882.

In *Titus Andronicus* we have an echo of the old classical imagery :—

> But straight they told me they would bind me here
> Unto tbe body of a dismal yew.—(ii., 3.)

THE ASPEN.

The aspen, *Populus tremula,* though at the present day in England by no means common, would seem to have been, in the Plantagenet times, abundant. For, in the fourth year of Henry V. (1417) the employment of the wood for purposes other than the manufacture of arrows was forbidden by Act of Parliament. Spenser, in the *Faëry Queene,* says it was "good for staves," whence, perhaps, the repeal of the law under James the First. Excelled by all our native foresters in length of existence, the aspen is still one of the most beautiful and interesting of British trees. The stature is inconsiderable; the flowers are inconspicuous; but the foliage is ample and refreshing. Individually the leaves are nearly circular. They have very long stalks, which are flattened laterally at the upper part, the blade quivering, in consequence, with the lightest breath of wind. Directly the wind stirs, see how delicate the ripple !— with noise, as it were, of a gently falling shower, though the branch, as a whole, does not stir. This charming peculiarity is observable also in the nearly allied *Populus nigra,* the common or old English "Black poplar," which latter tree is perhaps intended in certain references to the aspen, and *vice versâ*, or is at all events included therein. No wonder that in all ages the poets have made use of it when the subject for illustration has asked for a sensitive plant. It appears in Scripture in the memorable narrative in 2 Sam., v., 23, "When thou hearest the sound of a going in the tops of the *bekha'im*" (in the A. V. most unfortunately mistranslated "mulberry trees"), illustrating how the least of the phenomena of nature are no less immediately under the Divine control than the greatest and grandest, and recalling the other and equally suggestive picture of the "still small voice." Old Homer uses the flutter, not a leaf standing idle, as an image of the ceaseless whirling of the spindles by Alcinous' handmaidens (*Od.* vii., 106). Hypermnestra, in Ovid, tells us that just like this was the anxious beating of her heart (*Hyp. Lync.* 40). Chaucer adopts that beautiful line almost verbatim.

Shakspere knew the tree well. "Feel, master, how I shake! . . . Yea, in very truth do I, as 'twere an aspen leaf" (*2nd Henry the Fourth*, ii., 4). So in *Titus Andronicus*, ii., 5:—

> O, had the monster seen those lily hands
> Tremble, like aspen leaves, upon the lute.

The proper and actual name of the tree is simply "aspe," Anglo-Saxon æspe, as in Chaucer:—

> "As oak, fir, birch, aspe, alder, holm, poplere."
> *Palemon and Arcite.*

Aspen is the adjectival form, "tree" or "timber" being understood.

E

Though inconspicuous, the flowers of the aspen, as of all the other poplars, are remarkably pretty in structure. They are produced in the shape of catkins, long before the leaves expand, in the earliest of the Easter times,

> When winter, slumbering in the open air,
> Wears on his softened looks a dream of spring,—

differing, however, from most other catkins, in possession of abundance of a peculiar soft grey woolliness. For an amiable mind, that loves the little as well as the large, there is not a pleasure more thorough and unwearying than the examination of our British tree-flowers, so curious and delicate is the formation.

Chapter Fourth.

THE LINDEN.

In the line-grove which weather-fends your cell.

Tempest, v., 1.

HAKSPERE'S "line" is the beautiful tree which at present, though only since about 1700, is wrongfully called the "lime," which latter name belongs to an Indian Citrus, long celebrated for its fruit, and nearly related to the lemon. "Line" itself is not the original, being a shortened form of the Anglo-Saxon *lind*, which is connected in turn with *lentus*, pliant, and evidently refers to the usefulness of the inner bark as a material for string and cordage, very anciently recognized, and adverted to by Horace, who for some unknown reason prefers the Greek name of the tree to the Latin — *displicent nexæ philyra coronæ,* "chaplets woven with the rind of the line" (*Carm.* l., 38). The Shaksperean spelling, which some have thought to amend by alteration to the modern corrupted one, is vindicated in all the old herbals, and more than once in early verse, where it rhymes to *thine*—

> Now tell me thy name, good fellow, said he,
> Under the leaves of lyne.

"Linden" is the adjectival form of the word, "tree," being understood, thus corresponding with "aspen."

Less robust than the forest monarch, inferior in dignity to the elm, the linden is still entitled to count with the most delightful of English trees. It is the vegetable analogue of that happy condition of body which the ancient Greeks denominated εὐσαρκος, neither fat nor lean, but gracefully intermediate. It is one of the trees which go with the acacia and the birch in their representative character, conveying a certain elegant idea of feminine contour and attributes, as distinguished from the stalwart though less amiable masculine chestnut. Growing wild plentifully in woods in the form called by botanists *Tilia parvifolia,* in the improved one called *Europæa,* it has been from time immemorial a chosen ornament of parks and pleasure-grounds, and being often planted in avenues, has given new enrichment even to nature. The peculiarly good qualities consist, it is hardly necessary to add, in the symmetry of the outline, the lower branches often bending to the earth, so as to form a natural tent; in the remarkable ease and lightness of air and habit, this coming of the long stalks of the thin broad leaves; and in the abundance, in high summer, of the honeyed and fragrant bloom.

What other tree should Shakspere select to give shelter to the home of Prospero and Miranda? "The cell of Prospero," it has been remarked, "with its adjoining grove, is one of the most distinct and pleasing conceptions of natural scenery to be found in his works." Nestling under the lindens which "weather-fend" it, in this quiet island home we have not only seclusion and repose, the boughs disturbed only at times by the "light pinions of Ariel;" but that which gives to the *Tempest*— one of the only two of the romantic comedies produced by Shakspere which owes its plot to no previous author— its very marked and specially attractive interest, namely, the disclosure it affords of the writer's own personality. Everywhere else, Shakspere though present, is veiled; in Prospero we have him, not only beside us, but as the real and living man. Prospero is acquainted with all the secrets of nature; he penetrates men's minds and discerns their purposes. He has all the wise prevision, the authority and the gentleness of genuine power. He calls up magnificent visions; at his bidding "the air is filled with sweet music," or with the sounds of hound and horn; he raises or quells the storm; he commands, and a splendid banquet is spread.* Add to this the immortal words,

> My library
> Was dukedom large enough,

---

* Rev. J. Hunter. New Illustrations of Shakspere, i., 180.

and his infinite capacity for pourtraying the purest and
sweetest forms of human love—Miranda, gentle, affectionate, retiring, quite feminine Miranda, standing forth
as one of his most exquisite creations, and Shakspere
himself is veritably here—his powers, his temperament,
his genius, in perfect portrait. How sweet, too, the idea
of those beautiful trees in their supply of arbours and
shady alcoves! How often were they sought as shelter
from the noonday sunshine by the dear girl who,
when most enamoured, still "remembered" her "father's
precepts."

The honey of the flowers is not forgotten:—

Where the bee sucks, there suck I,

for surely this is the intent of the passage, quite spoiled
by the suggested change of the word to "lurk." Ariel,
when he has finished his tasks, desires nothing more
than to command it;—

Merrily, merrily, shall I live now,
Under the blossom that hangs on the bough.—(v., 1.)

In benevolence to the bees no tree wild in Europe
excels the linden. Were we not sure that Shakspere,
standing beneath, had many a time listened to their
drowsy murmur, we might think he had been lately
reading the story of the good old man in Virgil who so
loved his garden:—"Here planting among the shrubs,
white lilies, vervain, and esculent poppies, he equalled
in his contented mind, the wealth of kings. The first
was he to pluck the rose of spring, and the first to gather
the fruits of autumn; and even when sad winter split the

rocks with frost, and bridled the current of the streams with ice, yes, in that very season was he cropping the locks of the soft acanthus. Lindens had he, and pines, in great abundance; he, therefore, was the first to abound with prolific bees, and to strain the frothy honey from the well-pressed combs."—*(Georgic* iv., 131-141.)

One of the constituent trees of Prospero's grove appears to be spoken of in an earlier scene, when to be used for the display of the "glistening apparel" designed as a "stale" or bait for Caliban and his thievish companions. But the intent of the word is disputed, some of the best critics understanding it to denote a "clothes line," such as is used in washerwomen's drying-yards. Mr. Halliwell is of this opinion (vol. i., p. 39). It is supported also by Knight (Comedies ii., 440). But Shakspere would hardly mix up with poetry so beautiful as the picture of Prospero's home an idea so prosaic. If we are to think at all of the purpose of laundresses' clothes lines, it is quite as satisfactory, and incomparably more agreeable, as well as congenial to the time and place, to remember country-folks' employment of the hedgerow and the grass-plat. Further on, in the same scene, this particular "line," whatever it may be, is referred to, upon two occasions, jocularly.

THE HAWTHORN.

The "milk-white thorn" that in early summer dapples the hedgerows with its fragrant bloom—the sweet "May" that literally "scents the evening gale"—is, like the

linden, a genuine British plant, and when unmolested by
man, one of the most distinct and beautiful our island
possesses. It occurs in every part of Europe, and
extends even into Asia, but nowhere presents a lovelier
spectacle than with ourselves, the climate of England
being peculiarly favourable to its nature. The charms
increase, as with many other trees, with lapse of years;
the hold upon life is also remarkable—thorns count,
in truth, with the longævals, so that, however old, they
never look antiquated. When standing alone, as upon
the greensward of some ancient park, how beautiful the
large round head, the bright green of the opening leaves,
the snow of the innumerable blossoms, and in the fall,
when the foliage reddens, the emulation of the ruddy
fruit. Unequalled, in England, for "quick-set" fences,
the natural habit of the plant is in these, by much use of
the shears, entirely effaced. Hedges composed of it
have been in vogue from the earliest ages of civilization;
the name of "hawthorn" itself is a testimony thereto,
"haw" being no more than a modern shape of the
Anglo-Saxon word for a hedge: to Shakspere they were
as familiar as to ourselves;—almost always, when he
invites us to thought of this delightful old·plant, it is
with reference, however, to the hawthorn as an unspoiled
Tree. First, we have the cool and pleasant shadow,
always cast upon grass, the hawthorn, in the wild or
natural state, never hindering the growth of turf:—

. . . .

Ah ! what a life were this ! How sweet, how lovely!

Gives not the hawthorn-bush a sweeter shade
To shepherds, looking on their silly sheep,
Than doth a rich embroider'd canopy
To kings that fear their subjects' treachery?
O yes, it doth, a thousand times it doth.
And to conclude. The shepherd's homely curds,
His cold thin drink out of his leathern bottle,
His wonted sleep under a fresh tree's shade,
All which serene and sweetly he enjoys,
Is far beyond a prince's delicates,
His viands sparkling in a golden cup,
His body couchèd in a curious bed
When care, mistrust, and treason wait on him.

3rd Henry the Sixth, ii., 5.

In the *Midsummer Night's Dream* the hawthorn helps
to give that charming picture of the delicious season
marked not more by itself than by the anemones:—

Your tongue's sweet air,
More tunable than lark to shepherd's ear,
When wheat is green, when hawthorn-buds appear.

(i., 1.)

In this beautiful fairy-tale it supplies shelter also to the
would-be players:—

BOTTOM : Are we all met?

QUINCE : Pat, pat; and here's a marvellous convenient place for
our rehearsal. This green plot shall be our stage, this hawthorn-
brake our tyring-house.

How admirable, again, the indication of the season,
without naming it, in *King Lear:*—

Through the sharp hawthorn blows the cold wind.

(iii., 4.)

For in winter, the hawthorn, being a deciduous tree,

affords shelter no longer; and this not only because of its having cast its leaves—the under-structure is singularly thin and scanty; through the hawthorn, more than through any other tree of its kind, the wind whistles "to-and-fro conflicting," and there is not one that, when leafless, affords less shelter from the "pelting" of the "pitiless storm." The features in question are especially well marked in aged and wind-beaten trees upon bleak and desolate heaths and commons, thus giving the hawthorn a place incomparably just and true in the piteous scene in which we find it—the foremost figure the poor broken-hearted old man who has proved so bitterly

> How sharper than a serpent's tooth it is
> To have a thankless child !

Rosalind, as usual, brightens everything:—"There's a man haunts the forest, that abuses our young plants with cutting 'Rosalind' upon their barks; hangs odes upon hawthorns, and elegies upon brambles." Ah, for a glimpse of the heart-sunshine upon her countenance as she plucks them off—the light of her steadfast eyes as she peruses! The boughs are not snowy now, but aglow with berries, for it is autumn. "I found him," says Celia, a minute before, "under a tree, like a dropped acorn," her comparison suggested by the last object she has noticed on her sylvan way. In the entire drama there is not the slightest allusion to spring, spring productions, or spring phenomena. Everything mentioned has the scent and hue of ripe October. Rosalind's hawthorns, like the oaks, are heavy with fruit.

Drawn & Etched by W. Richardson

In one instance the citation of the hawthorn is meta-phorical, and like Benedick's of the willow, in *Much Ado*, upon the lines of the facetious. Coming from lively old Sir John, it could hardly be anything else: "I cannot cog, and say this and that, like a many of these lisping hawthorn buds that come like women in man's apparel, and smell like Bucklersbury in simple time."— *Merry Wives*, iii., 3. The picture of the "Verdant Greens" of the day, their mincing and affected talk, and their overdone self-perfumery, is perfect. Bucklersbury, a famous old London street, was in Shakspere's time noted for the number of its druggists, dealers, at that period, chiefly in odoriferous dried herbs—"simples," in the popular vocabulary. Why the knight should select "hawthorn-buds" rather than any other kind, as good for his scented foplings, does not, however, appear. It is a little singular, to say the least, that one of the old English names for these buds, when just expanding, is "Ladies' meat."

Mention of the hawthorn is scarcely ever made by the ancient or classical poets, and then only in reference to its supposed efficacy in averting the evil effects of unkind enchantments, intimidating demons, and healing snake-bites. Of the first-named superstition we have an instance in the *Fasti*, where a wand made of the wood is bestowed with a view to the protection of the sleeping infant from the red-jawed harpies (vi., 129, 165).

Chaucer makes amends in a well-known beautiful passage. Most indeed of the old English poets have something to say about the hawthorn, and this, generally in connection with the season of its bloom.

> Amongst the many buds proclaiming May,
> Decking the fields in holiday array,
> Striving who shall surpass in braverie,
> Marke the fair blooming of the hawthorn tree,
> Who finely cloathèd in a robe of white,
> Fills full the wanton eye with May's delight.

Remember, while reading these and all similar old English verses, that the year, in the time of Chaucer, Spenser, Shakspere, and onwards up to 1752, reckoned from twelve days later than at present, "May-day" being then what with ourselves is May 13th. All descriptions of spring and of the vernal charms of nature apply to a corresponding or twelve days' later period.

THE BOX-TREE.

Concerning other aboriginal British trees, Shakspere has very little to say, and the allusions, when they occur, have reference almost exclusively, to the economic uses. To introduce natural objects for purposes beyond simple illustration, or the giving of elegance to a picture, was no part, as already observed, of his design. There is no ground for surprise, accordingly, that what remains is of trifling amount. The box-tree, for example (if this be really British), being an evergreen, and extremely dense in foliage, becomes in *Twelfth Night*, ii., 5, a good hiding-place,—"Get ye all three into the box-tree." We must not think of it from the diminutive variety employed

in gardens as a fence or "edging" for flower-borders, nor
even from the bushy character attained in shrubberies.
When the circumstances of its long life are fairly congenial,
the box-tree is capable of attaining very considerable
dimensions. At Clifton Lodge, near Shefford, Bedford-
shire, there is one (unless since destroyed,) which, when
measured in 1865, was found to be close on twenty-two
feet high; the general spread of the branches was
twenty-six feet; and the trunk at a yard above the
ground, was fifteen inches in diameter. How much the
box was esteemed by the ancients as an ornamental
evergreen, hardly needs saying. Ovid refers to it in his
perpetuoque virens buxus (Met. x., 97). In another place
he well characterizes the abundant and close-set foliage,
densa foliis (A. A., iii., 691). Virgil speaks of the same
in that beautiful passage where he describes the hills as
undans, "waving," the epithet, which properly belongs to
the branches of the trees, being transferred from the
living thing to the inanimate one. When introduced into
the intensely artificial gardens of the ancient Romans,
this patient tree was subjected, like the yew, to that
odious clipping into grotesque and unnatural forms,
which in the Shaksperean age had become fashionable in
England, and of which there are memorials still extant.
The practice was consistent with the manners and
customs of a people who loved barbarities and the blood
of gladiators; but it is one from which all genuine and
cultured taste recoils, and with Shakspere, we may be
sure, it found no favour. Fine examples of the box in

its unmolested state, were in all likelihood, in the Shaksperean age, abundant in this country. The number of ornamental evergreens introduced at that early horticultural period was so limited, that they might be told upon the fingers. The laurels, the Mahonias, the rhododendrons of to-day, were quite unknown. In their absence the box would be esteemed to a degree we can now hardly imagine, and commensurate pride be taken in the spectacle of its green perfection. Chaucer takes up the Scriptural idea of the tree, or that which presents it to us as the emblem of Christian fortitude, but secularizes it into the simply imperturbable:—

> He like was to behold
> The box-tree, or the aspis dead and cold.
> *The Knight's Tale.*

THE HOLLY.

So with that supreme evergreen, old England's indomitable holly, the only wild one that shines—Shakspere, every Christmas-tide, admired its coral bracelets just as to-day we do ourselves. Branches of holly were employed in the Elizabethan period, by old usage, as the fitting symbol of radiant victory, life triumphing over death, glossy leaves and scarlet berries defeating frost and snow; thus of the great Advent which Christmas commemorates. Still he makes mention of it only once, in the little song in *As You Like It*, ii., 7. But who is the singer, and where is it sung? In Arden, by Amiens, who received his inspiration from its beautiful "green," and found in the presence of nature's wild holly abounding satisfaction.

THE BIRCH.

Notwithstanding its incomparable grace of figure, the delicate and unique whiteness of the stem, and the lightsomeness of the depending tresses—features which render it "the lady of the woods," Shakspere again speaks of the birch only once, and even then only for what it supplies. "I have not red," says old Turner, "of any virtue the birch hath in *physic.* Howbeit it serveth for many good uses, and for none better than for betynge of stubborn boys, that either bye, or will not learn."— (*Herbal,* 1551).

> Fond fathers,
> Having bound up the threatening twigs of birch,
> Only to stick it in their children's sight
> For terror, not to use; in time the rod
> Becomes more mock'd than fear'd.
>
> *Measure for Measure,* i., 4.

In attempting to mend obviously incorrect Shaksperean readings, we must take very particular care not to go yet further astray. It may be permitted, however, to ask, Is it possible that the birch can be the tree intended in the very perplexing passage in the *Tempest?*—

> And thy broom groves,
> Whose shadow the dismissèd bachelor loves,
> Being lass-lorn.—(iv., 1.)

Birch-twigs are used for the manufacture of besoms and "brooms" as well as for the stimulation of boys such as old Turner's, and Shakspere may have called the tree, in conformity with a very common custom of language, after the implement manufactured from it. Are we sure that he really wrote "broom"? In any case, that he meant the *Spartium Scoparium,* cannot for a moment be supposed. The capacities of the Spartium, and its dimensions, even when at the largest, forbid the idea. The birch, on the other hand, would serve the purpose perfectly. Slim in composition, offering the fewest impediments to free movement of any trees accustomed to grow in company, a more suitable retreat·than a grove of birches could hardly be offered to a purposeless, forlorn, and sauntering lover. "Shadow" does not necessarily imply a darkened covert. It is enough to understand, in the present instance, a place of seclusion, similar to Valentine's,

> This shadowy desert, unfrequented wood,
> I better love than flourishing, peopled towns.
> Here can I sit alone, unseen of any,
> And to the nightingale's complaining notes
> Tune my distresses, and record my woes.
>
> *Two Gentlemen of Verona,* v., 4.

THE ASH.

The ash, renowned, like the birch, for the singular beauty of its habit and profile, is also passed over by Shakspere, except in reference to the strength of spear-shafts made from the wood:—

O let me twine
Mine arms about that body, where against
My grainèd ash a thousand times hath broke.
Coriolanus, iv., 5.

No figure of speech is more common in language (as illustrated just above in the probable meaning of "broom"), and in that curious verse in Nahum, "The fir-trees shall be terribly shaken" (ii., 3), the sense being as here in Shakspere, the spear-shafts, which in the dreadful day foretold, are to be brandished aloft.

THE ELM.

The elm holds an agreeable place in Shakspere by reason of the beautiful image in the *Midsummer Night's Dream*;—

Sleep thou, and I will wind thee in my arms.
So doth the woodbine—the sweet honeysuckle—
Gently entwist—the female ivy so
Enrings the barky fingers of the elm.—(iv., 1.)

"Just as the woodbine and the ivy clasp and encircle the tree; so, my love," Titania means, will I, during thy slumbers, embrace *thee*." In *2nd Henry the Fourth* (ii., 4) a "dead elm" furnishes an odd kind of satirical metaphor; and in the *Comedy of Errors*, ii., 2, we have an allusion to the practice of the vine-cultivators of ancient Italy, who were accustomed to train their plants to young elm-trees, as many times spoken of in old Roman literature. The passage again represents the elm as masculine,—

F

> Thou art an elm, my husband, I a vine,
> Whose weakness, married to thy stronger state,
> Makes me with thy strength to communicate,

with the addition of a most delicately beautiful setting
forth of one of the loveliest instincts of woman,—that
one, which arising upon the strength and tenacity of her
affections, leads her always to cling to her mate, whose
own great pride is to render support. No one ever
preserved more thoroughly than Shakspere in pictures of
the relation of the sexes, the perfect and healthy balance
of manhood and womanhood which it pleased God to
design in the beginning.

THE ELDER.

The homely, old-fashioned elder of the hedgerow, and
of hillsides otherwise untenanted by an arborescent
plant, would become familiar to Shakspere in his boyhood, for is it not this to which every lad brought up in
the country resorts for toys and pop-guns? One can
easily imagine the recollection of the sports of his school-
days that would suggest the image in *Henry the Fifth*,—
"That's a perilous shot out of an elder-gun, that a poor
and private displeasure can do against a monarch"
(iv., 1); and that other in the *Merry Wives*, where the
pith, so light and compressible, so easily forced out,
stands for the reverse of "heart-of-oak,"—"what says my
Æsculapius, my Galen, my heart of elder?" (ii., 3). The
copious, creamy bloom of the tree, precisely concurrent

with midsummer, is somewhat honey-scented; the rich black-purple clusters of fruit recommend themselves for a kind of wine or cordial;—in such company we do not look for unpleasing scent of foliage; so much the more remarkable therefore the contrasted character adverted to in *Cymbeline*, where the tree supplies a metaphor quite seasonable:—

> And let the stinking elder, grief, entwine,
> His perishing root with the increasing vine.—(iv., 2.)

In *Love's Labour's Lost* the poet shows his acquaintance with the old tradition preserved by Sir John Mandeville, playing at the same time, upon the twofold meaning of the word;

> HOLOFERNES: Begin, sir, you are my elder.
> BIRON : Well followed : Judas was hanged on an elder.
>
> (v., 2.)

Lastly, in *Titus Andronicus*, the play one would gladly see dismissed altogether from the Shaksperean brotherhood, an elder marks the scene of a tragic incident;

> Look for thy reward
> Among the nettles at the elder-tree.
>
> .   .   .   .
>
> This is the pit, and this the elder-tree.—(ii., 4.)

THE SYCAMORE.

The sycamore holds a place intermediate between the trees, wild in our own island, of which Shakspere had distinct personal knowledge, and those which like the cedar, the cypress, and the myrtle, he talks of only from hearsay. Now universally diffused, and so thoroughly naturalized as to be included in the catalogues of indigenous British plants, in Shakspere's time the sycamore had been quite recently introduced from the mountainous parts of central Europe; and although he may have seen it in one of the "walks and places of pleasure of noblemen," where, according to Gerard, 1596, it was "specially planted for the shadow's sake," the probabilities are that he used the name purely by adoption. Every one has enjoyed the coolness given in summer by the abundant vine-like leaves, recalling, while in their green shade, the beautiful picture in the Æneid,

> ' Pinea silva mihi multos dilecta per annos,
> Lucus in arce fuit summâ, quò sacra ferebant,
> Nigranti piceâ, trabibusque obscurus acernis.*

Every one is prepared thereby for the corresponding picture in *Love's Labour's Lost*,

> Under the cool shade of a sycamore
> I thought to close mine eyes for half an hour—(v., 2),

and can appreciate, in equal measure, the preference felt for this charming tree by Juliet's lover;—

> Underneath the grove of sycamore
> That westward roveth from the city's side,
> So early walking did I see your son.
> *Romeo and Juliet*, i., 1.

---

* "Upon a lofty mountain stood a piny wood, by me through many years beloved, embowered with dark-hued firs, and the shady boughs of the sycamore, whither they brought me sacred offerings." (ix., 85-87.)

There is a trifle of lingering doubt, after all, whether while reading these two passages it would not be better to think of the Plane, the tree of immemorial honour with the ancients in respect of the delightfulness of the shade it offers, and certain to have been known to Shakspere by repute. To this day the sycamore has for its appellation with the scientific, "Pseudo-platanus," or "the mock plane." "Sycamore" itself denoted several different things—the scriptural one, a species of fig;— in Matthiolus, the melia; in Chaucer, in the *Flower and the Leaf,* some kind of scandent shrub, probably the honeysuckle. There is no reason however to doubt that with Shakspere, if, just possibly, not the *Acer Pseudo-platanus,* it would unquestionably be the yet nobler tree, *Platanus orientalis.* No tree, in its general character, is more imposing than the plane. There are loftier trees, and greener ones, and more flowery ones, but few present so large an aggregate of excellent qualities, these latter including a certain air of gentleness and repose, and a capacity for affording a peculiarly agreeable shelter from the heat of the sun, though no tree produces fewer leaves in proportion to the general plenitude of the foliage. The height is ordinarily about seventy feet; the boughs spread widely, but not in disproportion to the altitude; the leaves, individually, resemble those of the vine, but the clefts are much deeper, and the points are remarkably acute. The flowers are borne in globular clusters, these becoming spheres of brown seed, which dangle throughout the winter from the bare branches, as if ready and

waiting to drop to the ground, and present a singular
spectacle, the only one of its kind among forest trees.
The original species—the *Platanus orientalis*—appears
to have been introduced into this country by Lord
Chancellor Verulam.   Shakspere may have seen it.   In
any case the genuine sycamore has no greater claim
upon the poets.   Mention of the latter is made also in
Barbara's "song of willow" (*Othello,* iv., 3).

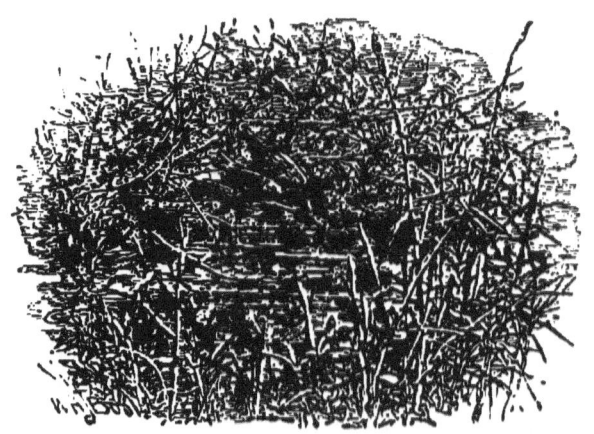

Chapter Fifth.

THE WILD-FLOWERS.

Honi soit qui mal y pense, write
In emerald tufts, flowers purple, blue, and white,
Like sapphire, pearl, and rich embroidery,

Merry Wives, v., 5.

S Shakspere's favourite trees were those of the woods in which as a boy he was accustomed to wander, absorbing "sweet influence," so are his wild-flowers those of the Warwickshire meadows, trodden, we may be sure, with equal delight. The green fields around his native village, the quiet lanes, the borders of the pretty streamlets carrying bubbles to the Avon, were their homes. Here it was that he first plucked the "pale primrose," the freckled cowslip and the early daffodil. To people who love him it is a source of perennial enjoyment that in these sweet old sanctuaries

of natural beauty, Shakspere's chosen wild-flowers still
abound, their original and simple loveliness unchanged.
Time, that deals so hardly with many stouter things,
is merciful to the fragile field-flower; everything that
Shakspere talks of abides in the old haunts, renewing
itself in beautiful annual flow. No part of our island is
richer than Warwickshire in genuine wild-flowers, so that
again it is a matter of thankfulness that Shakspere lived
not only *when* he did, but in his youthful days just *where*
he did; not only in the country, but in the very heart of
"merrie England," and in a region of peculiarly pleasing
self-adornment. He does not speak, as already said,
of any considerable number of different species. Including the two or three hedgerow flowering shrubs which
count, by a kind of prescriptive right, with the factors of
the wild garland, the honeysuckle to wit, and the
sweet-brier; including also some half-dozen of the inhabi-
tants of the border-land between weed and wild-flower
of which he makes mention, furze and heather for
example, the total is still barely a score. More, in truth,
are not wanted. The bouquet, as we have it, is unique,
and for perfection asks none besides. The charm of a
great poet's flower-scenery consists not in the abundance
or the variety of the objects mentioned, but in the
touches bestowed on a few, especially when those few
belong to the ranks of the "common." One of the most
trustworthy tests, perhaps the final test, of the great poet,
is that he makes the old-fashioned seem new, and the
unpretentious rare and golden; showing us, as he moves

along, that the most ordinary things of God's bestowal
are solar centres; giving life to the inanimate, and voice,
and even song, to that which previously was silent.
When this is done supremely, as by Shakspere, wise men
are content. There is a very remarkable difference also
in the degree of frequency with which he mentions these
twenty or so of different species. Of allusions to the
violet there are no fewer than eighteen examples; the
cowslip is noticed on six occasions; the orchis, the
harebell, the Lady-smock, and the dead-nettle appear, on
the other hand, but once—a matter, it may be, after all,
for congratulation, since when the instance stands alone,
the mind dwells upon it with so much the more interest
and curiosity. In one or two cases the application is
not clear. What "crow-flowers" are, and "cuckoo-
flowers," is still unproven ; probably they are not
intended to be taken as special names, but as generic
terms. Three considerations are always of great value
when there is uncertainty. Is the plant we believe to be
intended, one that grows naturally in Warwickshire, and
that Shakspere may in all likelihood have been familiar
with? Secondly, what light can we obtain from the old
herbalists, and from contemporary literature? And
lastly, as Shakspere never confuses the flowers of
different seasons, what company is it found in? That
"oxlip" and "wild thyme" stand side-by-side in a well-
known charming song is no objection, since we are not
required to suppose that the "bank" referred to was
made lovely by both at once. Shakspere's flowers

are in almost every instance those of the spring, though his weeds belong almost exclusively to summer and autumn.

THE VIOLET.

Over this immemorial and delicious little flower, the *Viola odorata,* so far as regards Shakspere, happily there can never be any demur. We have to thank him, note at the outset, not only for incomparable allusions to it, but for giving the name a fixity it never before possessed. Prior to the time of Shakspere, "violet" was one of the appellations borne by a dozen different flowers, some of which still retain it, as in the case of the "water violet," *Hottonia palustris,* the Dames' or "Damascus violet," *Hesperis matronalis,* and the Calathian violet, *Gentiana Pneumonanthe.* Fuchsius, in 1542, figures the snow-drop as "Viola alba:" Lobel describes the plant now called Honesty—that beautiful crucifer with the large oval silvery shields, so much valued, when dry, for parlour decoration, as Viola Lunaria major; the Can-terbury-bell was then Viola Mariana, the sweet-williams, Viola barbata. This wide diversity of application was by inheritance from the Romans, with whom viola was very seldom special; and more remotely, by inheritance from the ancient Greeks, whose ἴον, though it may have included the Shaksperean flower, most certainly meant many things besides, though the idea of purple seems to have been paramount, as when little Evadne, in Pindar, is said to be "ion-haired,"* having tresses, that is, of an impurpled raven-black. Is it nothing, then, to be glad of that the wanderer of nearly three thousand years at last should find a resting-place, and this the gift of Shakspere? What he would have done with the other "violets," had they called for mention, does not signify. Enough is it that he has fastened the name to the best of all. Shakspere, in so doing, further rendered the violet truly *his own*. When will language be as quick to thank the poets, who gave anchorage to names, as it is to honour the inventors of new arts and implements? Thus to commemorate Galvani, and Vernier, and Daguerre is most right and proper. Objects of natural beauty over which genius has cast her spell, are quite as worthy of appellations that shall remind us, whenever used, of the great and wise of past ages—the men who taught us how to see them, and how to appreciate their loveliness and significance. The day may yet arrive when instead of "the common daisy" it will be "Chaucer's daisy," and when people shall say, gratefully, not "the common violet," but "Shakspere's," perchance even "Perdita's." Shakspere was no stranger to the ancient celebrity of the name. The most charming of all the allusions to the flower he loved so well is that one in which he connects it with classical fable:—

> O Proserpina,
> For the flowers now, that frighted, thou let'st fall
> From Dis's wagon!

✓

---

* ἰοβόστρυχος, *Olympic,* vi., 30.

> Violets dim,
> But sweeter than the lids of Juno's eyes,
> Or Cytherea's breath!

What the myth intended, we do not know.   Ovid, who relates it, says they were violets and lilies—

> Quo dum Proserpina luco
> Ludit, et aut violas, aut candida lilia carpit.
>
> *Met.*, v., 391-2.

Neither did Shakspere know.   To him it was enough that it gave him a chance to dart at once into our souls, embedding old England's choicest wild-flower in the most romantic of primæval fairy tales.   Shakspere never imitates or copies for the mere sake of imitation.   One of the most delightful of his characteristics is that, without being a scholar, certainly without being up to the highest level of the scholarly age in which he lived, he still knew enough of the power and splendour of classical verse and classical story to have his own fancy enriched and stimulated by it.   When Shakspere copies or borrows, if either term can be justly applied to the use he makes of his predecessors, he repays a hundredfold in the method of the use.   Reading in ancient dress the wild old poetic fable to which he adverts, he takes just so much of it as will allow of rejuvenescence in the brimming founts of his own imagination.   Perdita's are not Ovid's flowers, for those may be anything we like to fancy: hers are old England's violet, and nothing besides.   These it is which Shakspere gives us; illustrating once again that other grand quality of the work of the master whereby we feel when in his presence that we do not so much hear about things, or read about them, as possess them.

Whether Shakspere was master enough of Latin to go to the original for this and other tales it must be admitted is an open question. In any case he would find the story in one of the many reprints of Arthur Golding's translation of the Metamorphoses, originally published in 1567. Ovid, very naturally, was one of the earliest of the old classical authors to receive translation, several besides Golding trying their hands. But what does it matter, even could it be demonstrated that he was wholly ignorant of any language except his own? That which is precious in Shakspere, and of incomparably greater importance to us, is the outcome of his own genius, the honey of his own teachings.

Similar, one cannot but think, is the Shaksperean indebtedness, which yet is no debt, in regard to that most sweet and tender passage in the *Fasti*, where the poet enjoins us to pay kindly respect to the tombs of the departed by strewing flowers upon them. "They ask," he says, "but humble offerings. To them is affection more pleasing than a costly gift. Enough for them are chaplets . . . and plucked violets" (ii., 539). For almost like an echo of these lovely words are the lines in *Pericles*, where Marina enters with her basketful of mourning tribute:—

No, no, I will rob Tellus of her weeds,
To strew thy grave with flowers: the yellows, blues,

> The purple violets, and marigolds,
> Shall, as a chaplet, hang above thy grave
> While summer days do last.--(iv., 1.)

The resemblance to the immortal passage in *Cymbeline*—

> While summer lasts, and I live here, Fidele,
> I'll sweeten thy sad grave,

may be accepted as a proof that, however little else
there may be of Shakspere's in *Pericles*—a play not
included in the first folio—Shakspere is at all events
present *here*. In any case the violets are again English
ones, violets definitely, because Shaksperean.

An emendation seems to be needed in the line,

> Shall, as a *chaplet*, hang above thy grave,

for although chaplets of the kind so called to-day, were
undoubtedly used in Shakspere's time, in funeral ceremonies, as illustrated in *Hamlet*, v., 2,

> Yet here she is allowed her virgin crants,
> Her maiden strewments, and the bringing home
> Of bell and burial,

the sense is that which elsewhere in Shakspere is
expressed by "carpet." "Carpet," in all likelihood, was
the word originally written. Carpets, in the modern
sense of the word had not, in the Shaksperean times,
come into use. Their place was held by rushes; and
"carpet" was then the name of a richly embroidered
table-cloth, as in the *Taming of the Shrew*, iv., 1:—
"Is supper ready, the house trimmed, rushes strewed?
Be .   . the carpets laid, and everything in order?"
A beautiful "carpet" of this character, and not a

"chaplet," is what Marina proposes to place upon the grave, the green grass of which is inlaid by her with the flowers. This is the view taken of the passage by Mr. Chas. Knight.* Dr. Schmidt, in the Shakspere lexicon, also approves it, placing the passage alphabetically, not under "chaplet," but under "carpet." That Dionyza, a minute afterwards, says to Marina, "Give me your *wreath* of flowers," in no degree sustains the "chaplet" hypothesis. The wreath would be brought, as in *Hamlet*, not as the sole and exclusive offering, but as supplementing the "strewments." The established practice in the Shaksperean times was plainly to scatter flowers, as from a basket, as illustrated in *King Henry the Eighth*,

> When I am dead, good wench,
> Let me be used with honour ; strew me o'er
> With maiden flowers, that all the world may know,
> I was a chaste wife to my grave.—(iv., 2.)

Again in *Cymbeline*, iv., 2, and again in *Romeo and Juliet;*

> Sweet flower, with flowers I strew thy bridal bed,
> Sweet tomb, that in thy circuit dost contain
> The perfect model of eternity.—(v., 3.)

---

* "*Carpet.* So the old copies. The modern reading is *chaplet*. But it is evident that the poet was thinking of the green mound that marks the last resting-place of the humble, and not of the sculptured tomb to be adorned with wreaths. Upon the grassy grave Marina will hang a *carpet* of flowers—she will *strew* flowers, she has before said. The carpet of Shakspere's time was a piece of tapestry, or embroidery, spread upon tables ; and the real flowers with which Marina will cover the grave of her friend might have been, in her imagination, so intertwined as to resemble a carpet, usually bright with flowers of the needle."

"He came," says the little page, giving an account of all that happened that mournful night,

> He came with flowers to strew his lady's grave.

"Crants," a word found nowhere else in English litera-ture, is a varied spelling of the German and Danish "cranz," a garland. That it should occur in the same play as *hebenon*, is at least a curious coincidence.

Similar yet again, one cannot but think, is the echo of classical verse when Laertes says in *Hamlet*,

> Lay her in the earth,
> And from her fair and unpolluted flesh
> May violets spring. I tell thee, churlish priest,
> A minist'ring angel shall my sister be,
> When thou liest howling.—(v., 1.)

For Persius, speaking of a deceased poet, says:—

> Nunc non cinis ille poëtæ
> Felix? non levior cippus nunc imprimit ossa?
> Laudunt convivæ: nunc non e manibus illis,
> Nunc non e tumulo fortunataque favilla
> Nascentur violæ?—(i., 36-40.)

We must be careful, at all times, never to over-estimate Shakspere's particular knowledge, whether of nature, of science, or of books—we must be quite as careful also never to confound simple coincidences with adaptations. Persius may have been read by Shakspere; many editions of the famous satirist had already been published; the volume, in an age so literary as the Elizabethan, was unquestionably within his reach; Ben Jonson distinctly quotes from the prologue in the *Poetaster;* it is quite possible, therefore, that the Hamlet imagery may have been suggested by the Roman:—note, however, how much we gain once more through Shakspere having made the appellation specific. Persius employed *viola* in the same broad, collective, indefinite sense which it holds in Ovid, as proved by a later passage (v., 182). In Shakspere we have the violet *ipsissima*, and the lines before us, by its introduction, acquire all the freshness and the fragrance of the morning.

The inmost sense of Laertes' words is that the pure and chaste, though it may be veiled awhile, is inextinguishable. The violet is introduced as their offspring, because the emblem, in the middle ages, and probably ever since the time of the Troubadours and the Crusaders, of faithfulness, constancy, loyalty, everything summed up in that beautiful old Greek word εὐστάθεια—

> Violet is for faithfulnesse,
>   Which in me shall abide ;
> Hoping likewise that from your heart
>   You will not let it slide,
> And will continue in the same,
>   As you have nowe begunne ;
> And then for ever to abide,
>   Then you my heart have wonne.*

---

* *A Handefull of Pleasant Delites*, 1584. One of the popular song books of the first half of the reign of Elizabeth, originally published in 1566. The present stanza is the seventh of fifteen constituting "A Nosegaie alwaies sweet, for Louers to send for Tokens of loue, at Newyeres tide, or for fairings, as they in their minds shall be disposed to write." Shakspere was certainly acquainted with it.

G

Ophelia herself has adverted to it in that intensely mournful utterance, while distributing her flowers among the company around—"I would give you some violets, but they withered all when my father died" (iv., 5). The poet's highest function, as we all know, is not so much to describe nature as to *insoul* it. How supremely beautiful in the present instance the imparting to the little wild-flower the deepest and saddest of human emotions; it perishes overwhelmed by grief and dismay.

But we have not yet done with Perdita. Why "violets dim?" Why "sweeter than the lids of Juno's eyes?" and yet again, than "Cytherea's breath?" The eyes are the acknowledged "windows of the soul." Shakspere transfers the name, by permitted metonymy, to the *lids*, a favourite usage with him, for we have it in *Romeo and Juliet:*—

> The roses in thy lips and cheeks shall fade
> To paly ashes; thine eyes' windows fall
> Like death, when he shuts up the day of life.—(iv., 1.)

In *Venus and Adonis:*—

> Her two blue windows faintly she upheaveth.

And again in *Cymbeline*, ii., 2:—

> Under these windows, white and azure-laced
> With blue of heaven's own tinct,*

in which last exquisite picture we are reminded that when the skin is exceptionally white and translucent, the

---

* Often wrongly punctuated, "white and azure, laced," etc. Azure, instead of being followed by a comma, is part of the obvious compound azure-laced.

veins show like a delicate faint blue net-work. The larger and rounder the eye, the more beautiful, of necessity, the azure lace. Eyes and eyelids of exactly this description were possessed by the most celebrated woman of Shakspere's own period—not of royal birth— Elizabeth Hardwicke, Countess of Shrewsbury, the irresistible though unloveable lady whose fourth husband was custodian, during many years, of the ill-fated Mary "Elizabeth Hardwicke," says her biographer,* "had one of those delicate complexions through the transparency of which the violet veins gleam beneath the skin in tender waving threads, giving a peculiarly beautiful tint to the forehead and eyelids." Mark the word "tint," Shakspere's own. Network produced by some deeper hue, absolute network, is reserved for the henbane and the pencilled geranium. This he does not mean. It never occurs in the violet, which in the white variety, so elegantly pointed to in the epithet "dim," † is, nevertheless often suffused with blue. Whatever the prevailing colour of the petals, the streaks which spring from the centre show plainly, as referred to particularly in *Venus and Adonis:*—

> These blue-veined violets whereon we lean,
> Never can blab, nor know not what we mean.

Elizabeth Hardwicke's countenance was very possibly not unknown to him, and before his mind. In any case

---

* Miss Costello, Memoirs of Eminent Englishwomen, vol. i., p. 10.
† Is it possible that Shakspere here gives new life to the "pallentes violas" of Virgil, Ecl. ii., 47?

he knew that Juno, queen of heaven, was not only
majestic in figure and stature, but distinguished for the
beauty and the prominence of her great eyes, the lids of
which must have been ample indeed. Homer calls
her βοῶπις, "ox-eyed," an epithet not altogether pretty
perhaps, in prosaic view; enviable, nevertheless, and
varied into the name of a celebrated radiant flower—
the ancient Buphthalmum, the modern "ox-eye daisy."
"Well," says Leigh Hunt, "does it suit the eyes of that
goddess, since she may be supposed, with all her beauty,
to want a certain humanity. Her large eyes look at you
with a royal indifference—Shakspere has kissed them,
and made them human, shutting up their pride."*

Every man knows of a voice that is sweeter than all
music; yea, of a footfall that is melody; he knows, too,
where to find fragrance sweeter than flowers can yield:—

> The forward violet thus did I chide :
> Sweet thief, whence did thou steal thy sweet that smells,
> If not from my love's breath?—*Sonnet*, xcix.

So in *Cymbeline:*—

> The leaf of eglantine, whom not to slander,
> Out-sweetened not thy breath.

What the lover finds in his betrothed, the ancients
thought of their goddesses:—

> Non Beroe vobis, non hæc Rhœteïa, matres,
> Est Dorycli conjux : divina signa decoris,
> Ardentesque notate oculos : qui spiritus illi,
> Qui vultus, vocisve sonus, vel gressus eunti !
>
> *Æneid*, v., 646.

---

* Men, Women, and Books, p. 149.

Iris is spoken of here, but the same attributes were given to Venus, many times in the Æneid called Cytherea, since it was near the island of Cythera, adjacent to the Peloponnesus, that, according to the ancient fable, she rose out of the smiling sea;—on two other occasions so called also by Shakspere,

> And Cytherea, all in sedges hid,
> Which seem to move and wanton with her breath,
> > *Taming of the Shrew*, Induction.

> Cytherea !
> How bravely thou becom'st thy bed, fresh lily !
> > *Cymbeline*, ii., 2.

By "Cytherea's breath" we are thus to understand the choicest imaginable of odours, surpassed in sweetness, nevertheless, by the breath of old England's little wild-flower.

Allusions to the perfume of the violet occur in several admired passages:—

> Therefore, to be possessed with double pomp,
> To guard* a title that was rich before,

---

* "To guard," used in a sense now obsolete, though many times met with in Shakspere, the sense being first to protect the edge of anything, as the garrison does the castle-wall; then, as defence is always ornamental, to adorn the edges of articles of dress with lace or other decorative material; then, metaphorically, to glorify a man's reputation, character, or "title." Thus, in the *Merchant of Venice*, ii., 2. :—

> Give him a livery
> More guarded than his fellows.

So in *King Henry VIII.*, Prologue:—

To gild refinèd gold, to paint the lily,
To throw a perfume on the violet;
To smooth the ice, or add another hue
Unto the rainbow, or with taper-light
To seek the beauteous eye of heaven to garnish,
Is wasteful, and ridiculous excess.—*King John*, iv., 2.

That strain again !   It had a dying fall;
O, it came o'er my ear like the sweet south
That breathes upon a bank of violets,
Stealing and giving odour.—*Twelfth Night*, i., 1.

The sweet south, in the *Twelfth Night* verses, though very generally accepted, is one of the doubtful phrases still in dispute. In all the early editions of Shakspere the words are printed "like the sweet *sound,*" the emendation having been made by Pope. Those who contend that in the manuscript it must have been "sound," point to other passages in which Shakspere represents the south wind as the very contrary of "sweet." In *As You Like It*, for instance—

You foolish shepherd, wherefore do you follow her,
Like foggy south, puffing with wind and rain ?—(iii., 5.)

It is further contended that if he really wrote "south," it would not be as signifying the south wind, but in the sense of "sough," used provincially to denote the "soft susurrus of the breeze." Pope's proposed correction is thought to be sustained by a passage in the *Arcadia* of

A fellow
In a long motley coat, guarded with yellow.
When in *Much Ado*, i., 2, Benedick says to the Prince of Arragon, "The body of your discourse is sometimes guarded with fragments," he means that the royal conversation is sometimes frivolous.

Shakspere's contemporary, Sir Philip Sidney:—" Her breath is more sweet than a gentle south-west wind which comes sweeping over flowery fields and shadowed waters." Perhaps Shakspere may have had in his mind that beautiful verse in the Song of Solomon—" Come, thou south wind; blow upon my garden, that the odours thereof may flow out" (iv., 16). "South," rather than "sound" is acceptable also from its bringing the passage into agreeable harmony with another in *Cymbeline,* quite as familiar:—

> O thou goddess,
> Thou divine Nature, how thyself thou blazon'st
> In these two princely boys ! They are as gentle
> As zephyrs, blowing beneath the violet,
> Not wagging his sweet head; and yet as rough,
> Their royal blood unchaf'd, as the rudest wind
> That by the top doth take the mountain pine,
> And make him stoop to the vale.—(iv., 2.)

In two places the violet is employed metaphorically:—

> Welcome, my son ! Who are the violets now,
> That strew the green lap of the new-come spring ?
> *Richard the Second,* v., 2.

"Who are the courtiers, that is, who attend on Boling-broke, now in the spring-time of his reign?" The other instance is in that profoundly moving scene, in pathos and delicacy of finish unsurpassed anywhere in Shakspere, where poor dove-like Ophelia, most *spirituelle* of all his conceptions of feminine grace, stainless as the pearl of the deep sea, fragile as the blue flax blossom, capable herself only of innocence and celestial trustful-

ness, and who never told her love, is warned by her
brother not to believe too confidently :—

> For Hamlet, and the trifling of his favour,
> Hold it a fashion, and a toy in blood;
> A violet in the youth of primy nature,
> Forward, not permanent, sweet, not lasting,
> The perfume and suppliance of a minute—
> No more !
> OPHELIA : No more but so ?
> LAERTES : Think it no more !

Still she is unconvinced; woman's faith, where the heart
has once gone over, is impregnable; her father himself
scarcely makes deeper impression :—

> OPHELIA : My lord, he hath importun'd me with love
>   In honourable fashion.
> POLONIUS: Ay, fashion you may call it. Go to ; go to.
> OPHELIA : And hath given countenance to his speech, my lord,
>   With almost all the holy vows of heaven.
> POLONIUS: Ay, springes to catch woodcocks.

The end we all know.   There is no need to go further
towards it at present.   Compare, rather, the tender lines
in Sonnet xii. : —

> When I behold the violet past prime,
> And sable curls all silver'd o'er with white,
>
> .   .   .   .
>
> Then of thy beauty do I question make,
> That thou among the wastes of time must go,
> Since sweets and beauties do themselves forsake,
> And die as fast as they see others grow.

"Prime" and "primy," in the above passages, signify
of or belonging to the Spring, as again in Sonnet xcvi. :—

> The teeming autumn, big with rich increase,
> Bearing the wanton burden of the prime.

" Forward," in turn, means " early," as in Sonnet xcix., above quoted. " Favour" means appearance or complexion, thence the countenance, as in many other places.

> My life upon't, young though thou art, thine eye
> Hath stay'd upon some favour that it loves.
> *Twelfth Night*, ii., 4.

So in *As You Like It :—*

> The boy is fair,
> Of female favour, and bestows himself
> Like a ripe sister.—(iv., 3.)

And in *Pericles :—*

> Ah, how your favour's changed !—(iv. 1.)

The remaining allusions to the violet, making the total of eighteen, occur in the *Midsummer Night's Dream,* ii., 2, in Oberon's little song, "I know a bank," upon which, in its season,

> The nodding violet grows ;

in *Cymbeline,* i., 6,

> The violets, cowslips, and the primroses,
> Bear to my closet ;

in *King Henry the Fifth,* iv., 1, "I think the king is but a man as I am ; the violet smells to him as it doth to me ;" in *Measure for Measure,* ii., 2,

> It is I,
> That lying by the violet, in the sun,
> Do as the carrion does, not as the flower,
> Corrupt with virtuous season ;

in *Love's Labour's Lost*, v., 2,

> When daisies pied, and violets blue,
> And cuckoo-buds of yellow hue,
> And Lady-smocks, all silver-white,
> Do paint the meadows with delight,

where, as violets, definitely so called, never grow in the open "meadows," we are to understand, by the latter word, in regard to these flowers, the country in general; and, on a second occasion, in *Venus and Adonis*,

> Who, when he lived, his breath and beauty set
> Gloss on the rose, smell to the violet.

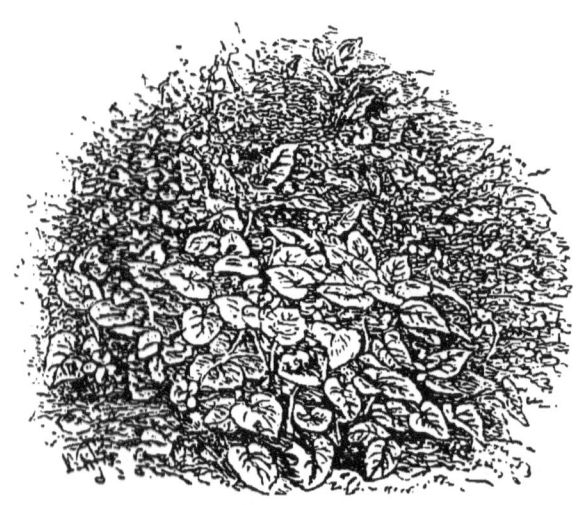

Chapter Sixth.

THE PANSY.

My crown is in my heart, not on my head;
Nor decked with diàmonds and Indian stones,
Nor to be seen:—my crown is called Content.
3rd Henry the Sixth, iii., 1.

HE pansy, *Viola tricolor*, is, like the violet, a genuine English wild-flower, coming up abundantly in dry waste places, especially where the soil has been ploughed, and blooming freely from April till October. There can be no doubt that it was one of the first to be brought into the garden, where the form and colours would soon improve. All the early botanists make mention of it as a well-known plant, some giving rude woodcut representations, as Matthiolus, in the very curious *Epitome de Plantis*, printed at Florence in 1586, in which volume it is called

Flos Trinitatis. The author adds that the French name
is "pensée," whence, very obviously, the Shaksperean
one, though when first employed in our own country
there is no evidence to show. That continental names
should accompany flowers brought as exotics across the
Channel, as in the case of the sweet-william, originally
the sweet œillet (of which latter word "william" is a
corruption), would be next to a matter of course. It is
curious, however, that an indigenous flower, named
already by the Anglo-Saxons, should receive, as in the
present instance, a new one of foreign birth. Why called
by a name so poetical is likewise unknown, the tradition
not having come down, though it may reasonably be
supposed to have been extant in the Elizabethan age.
The derivation, as of so many other picturesque names,
belongs to that beautiful department, not of our powers,
but of our inability, which is concerned with the best
and most precious knowledge. To find reasons for the
prosaic feelings and actions of life, to understand low
and vulgar things, is easy enough: ask the heart, on the
other hand, why it loves, why a dozen years of absence
make no difference to an inmost affection, and it is
mute ;—it knows only that faithfulness is a kind of
sunshine, and, like mercy, "twice blessed,"

> It blesseth him that gives, and him that takes.

Just so with the richest part of language ; there is a
romance in it to which the dictionary gives no index.
That some sweet and pretty superstition or fabled

attribute lies within the old French name, derived, it would seem, etymologically, from *pensare*, to weigh (in the mind), whence also "pensive," we may be sure, in truth, from the use made of it in *Hamlet;* for Shakspere, though fond of a verbal conceit, would never place one upon the lips of poor distracted Ophelia. The passage again counts with the most touching in his works. The king, the queen, Laertes, and various nobles are in conversation ; suddenly she enters, "dressed fantastically with straws and flowers," present in all her loveliness of form and feature, but her mind——?

> LAERTES : O heat, dry up my brain ! Tears seven times salt,
> Burn out the sense and virtue of mine eye !
> By heaven, thy madness shall be paid by weight,
> Till our scale turn the weight. O rose of May !
> Dear maid, kind sister, sweet Ophelia !
> O heavens ! is't possible, a young maid's wits
> Should be as mortal as an old man's life ?
> Nature is fine in love, and while 'tis fine,
> It sends some precious instance of itself
> After the thing it loves.

Ophelia sings; she utters she knows not what; then looking round, with a gleam of returning intellect, distributes her little gifts. Her soul is with Hamlet, far away, for whom she possibly mistakes Laertes, and she begins with talking to *him*. "There's rosemary, that's for remembrance; pray you, love, remember." Rosemary was one of the plants always carried at weddings; the dream of her own bridal is uppermost, but how exquisite the reference without mention of the word ! Then comes the soft, sweet feminine adjuration—look at her eyes—"I pray you, love, remember;" followed by "and there is pansies, that's for thoughts." "Thoughts?" Here one is constrained to pause, wondering if the pleasant fable can be pointed to in any other of the half-dozen old-fashioned appellations alive still in country places, such as "kiss me," most of which seem to refer, like the violet, in its emblematic character, to the idea of fond attachment, the very root and pivot of which is *pensez à moi.* Probably in the middle ages the pansy and the violet, flowers of extremely near botanical affinity, would be at times confounded or interchanged. There is no evidence as to the origin of the association, and we have simply to accept it as it stands. Laertes' expression, "Nature is *fine* in love," rests upon the curious lateral signification of the word,—in colloquial usage more familiar in the derivative *finesse,* meaning subtilty of contrivance, great "nicety," of artifice. In *All's Well,* v., 3, the king says to *Parolles,* "But thou art too fine in thy evidence; therefore stand aside." In Bacon's Apophthegms, it is used in the complimentary sense— "Your Majesty was too fine for my Lord Burghley."

The pansy is mentioned many times by the writers of the Shaksperean age, but scarcely two of them agree as to the shape of the word. Spenser, the last minstrel of chivalry, calls it, in the *Shepherd's Calendar,* the "pretie pawnce." Ben Jonson has

> I pray, what flowers are these?
> The panzie this;
> O, that's for lovers' thoughts.

In William Webbe's *Discourse of English Poetrie,* 1586,
in the course of his translation of Virgil's second eclogue,
we have "pancyes." Still another form occurs in Lyte.
"The pances or hartes ease,"* he tells us, is also called
"love in idleness."

These various spellings of a name so familiar in the
Elizabethan age, show how little stress is ever to be
laid upon any spelling of the period. A standard of
orthography did not then exist. Authors spelt very
much as they liked, and frequently by mere guess. Conformity was scarcely attainable even by the learned, and
when a manuscript was placed in the hands of the
compositors, if it did not find them ignorant and incom-
petent, it was sure of their carelessness. The men who
set the types of the early editions of the Shaksperean plays
were not only negligent themselves to a degree which in
modern times is astounding; but had no "reader" to
set them right before the printing was commenced. It
would not be difficult to find examples of several
different spellings often in a single play. So far as
regards "pansy," the consideration is of no importance.
It acquires great weight, however, when the question is
raised as to what spelling should we employ to-day when
reprinting. For philological purposes, the old is doubt-

---

* The name of "heartsease," it may be remarked (not occurring
in Shakspere), pertained originally to the wallflower, as may be
seen in Turner, ii., 163, and became transferred as an *alias* to the
pansy through an odd misunderstanding, explained at length by
Dr. Prior.

less very serviceable; but people in general go to Shakspere for his poetry, his art, his pathos, his infinite humour, his inexhaustible humanity, and these they gather best from the spelling they are used to. Webbe, it may be added, places the pansy in Virgil without authority in the original. No distinct allusion to it occurs in the classical poets, the flower mentioned in the sixth Olympian, 55, thought by some to be the *Viola tricolor*, being far more probably, as shown by the context, the yellow water-flag, *Iris Pseud-acorus*, the petals of which, as the author says, are purple-streaked. Pansies would never so mix with waterside reeds as to supply a place of concealment for a little child. The "luteæ violæ" of Catullus, xix., 12, cannot be adduced with any greater show of reason.

On two occasions out of the three on which Shakspere refers to the pansy he calls it by the other old name, just quoted from Lyte, "love-in-idleness," one which points anew to some unrecorded myth or superstition. The name is considered by the etymologists to be an elongation of "love-in-idle," loving, that is to say, fruitlessly, to no purpose, in vain, these having been the primitive senses of the term, as in *King Lear*, iv., 4 :—

> Darnel, and all the *idle* weeds that grow
> In our sustaining corn,—

and very distinctly, many times, in the early versions of Scripture. He prefers it now because the occasion is mirthful, as when Lucentio says, in the *Taming of the Shrew*, thinking about Bianca :—

But see, while idly I looked on,
I found the effect of love-in-idleness.—(i., i.)

What that effect was is told in the *Midsummer Night's Dream*, almost the whole of the entertaining circumstances of which intensely poetical drama, unique of its kind, are brought about by the magical virtues of the juice. But do not let us go too fast. First there is the beautiful allegory which, leading up so gently to the little flower, and before we have any idea of what is to come, gives it an interest almost royal. Oberon, king of the fairies, calls to his private and confidential page :—

> My gentle Puck, come hither : Thou remember'st
> Since once I sat upon a promontory,
> And heard a mermaid on a dolphin's back,
> Uttering such dulcet and harmonious breath,
> That the rude sea grew civil at her song,
> And certain stars shot madly from their spheres
> To hear the sea-maid's music?
> PUCK : I remember.
> OBERON : That very time I saw (but thou could'st not),
> Flying between the cold moon and the earth,
> Cupid all arm'd : a certain aim he took
> At a fair vestal, thronèd by the west,
> And loos'd his love-shaft smartly from his bow,
> As it should pierce a hundred thousand hearts :
> But I might see young Cupid's fiery shaft
> Quench'd in the chaste beams of the watery moon,
> And the imperial votaress passed on,
> In maiden meditation, fancy free.

So far the charming introduction, designed, as every one knows, to fix the thoughts upon Queen Elizabeth, the maiden monarch, whose love of elegant compliment, yea,

H

even of flattery, not to say adulation, Shakspere knew
full well; and, skilful courtier as he was, lost no opportunity of gratifying; and who yet, if history be true, was
not at all times "fancy free." For we are not to interpret
this beautiful phrase by the meaning of "fancy" current
to-day—some sportive conception of the mind, not
always reasonable, more often whimsical, and often connected with something ignoble. In the Shaksperean age
"fancy" was synonymous with "love," and perhaps
not unadvisedly, since the preferences felt by men and
women for one another admit of no explanation, and
belong purely to the realm of Taste. Compare, for
example, in iv., 1, of the same play,

> Fair Helena, *in fancy* following me.

So, in *As You Like It*, iii., 5, where Silvius presses his
love-suit upon the unwilling shepherdess ;—

> Sweet Phœbe, do not scorn me ; do not, Phœbe :
> Say that you love me not, but say not so
> In bitterness.
>
> .　　　.　　　.　　　.　　　.
>
> 　　　O dear Phœbe,
> If ever (as that ever may be near),
> You meet in some fresh cheek the power of fancy,
> Then shall you know the wounds invisible
> That love's keen arrows make.

Little, he means, as you may to-day conceive of the
force of love, you will learn all when it comes to your
own turn to be smitten by a countenance that is song
and sunshine. Juliet's attention is first attracted towards

Romeo by seeing him "fancy sick and pale of cheer"
for love of a cold beauty. In the *Midsummer Night's
Dream* Hermione calls wishes and tears "poor fancy's
followers." In *As You Like It*, again, reflecting idly
upon love, is

> Chewing the food of sweet and bitter fancy.—(iv., 3.)

Lastly, to cite no more of twenty instances, in *Troilus and
Cressida*, v., 2, the word is used as a verb :—

> Never did young man fancy
> With so eternal and so fixed a soul.

The introduction concluded, then comes, like a new and
glorious wave of the fast-flowing tide upon the beach,
the picture even more beautiful :—

> Yet marked I where the bolt of Cupid fell:
> It fell upon a little western flower,—
> Before milk-white, now purple with love's wound,
> And maidens call it "love-in-idleness:"
> Fetch me that flower. The herb I show'd thee once,
> The juice of it on sleeping eyelids laid,
> Will make a man or woman madly dote
> Upon the next live creature that it sees.
> Fetch me this herb, and be thou here again,
> Ere the leviathan can swim a league.
>   PUCK: I'll put a girdle round about the earth
> In forty minutes.—(ii., 2.)

The scene of the play being laid in Athens and a
neighbouring wood, it is allowable to regard the latter as
Oberon's home. We breathe, in truth, while reading it,
the atmosphere of Arcadia, sweetened by the mild waft,
never ceasing, from the blue Ægean; for Shakspere's

"fairies" are no other than the wood-nymphs of classic
fable, the old, old story in varied recital. The "fairies,"
in truth, beloved of the innocent fancy, are the same all
the world over, and in every age. The scene being
Athenian, hence, in turn, come the beauty and grace of
the little epithet "western," which, taken in connection
with the reference to Elizabeth, at once transports the
mind to Old England.

It is by delicate touches such as these that the great
poet is declared. Like the true artist, in his details he
is always a mathematician, as strong in the simplest
particulars as in the weightiest; and the word that seems
the most trivial not uncommonly carries as much meaning as all the others put together. Shakspere, it has
been well said, never plays in the shallows, but goes
down into the depths, speaking not only to our knowledge, but to our spirits. To Greece the little pansy of
England is emphatically a "western" flower. Saying
so, we are refreshed, in the most charming manner, by a
reminder of Home.

How elegant, too, the added fiction of the change of
colour, an ancient fable newly told. When in company
with Titania, and in fairyland, we must not insist
upon a too literal exactitude as to facts. That one of
the wild forms of the *Viola tricolor*, called by botanists
the *Viola arvensis*, has flowers white in every part excepting the golden recess in the middle, streaked with purple,
is quite true; it is true also that the petals are capable
of becoming purple under cultivation. Shakspere does

not refer to *this*. He transfers to the pansy what in
Ovid he had read about the mulberry, white as snow up
to the moment when "in such a night"—

> Did Thisbe fearfully o'ertrip the dew,
> And saw the lion's shadow ere himself,
> And ran, dismayed, away,

but presently, and ever afterwards, the colour of the
blood-stained veil.

The efficiency of the juice, need it be said, is amply
proved, and in twofold case. The fun is supreme; it is
a relief, nevertheless, to find that Oberon knows of
"another herb," an antidote that will "take the charm
from off her sight," and that Titania, with all her married
wilfulness, is not to be punished indefinitely :—

> Be as thou wast wont to be;
> See as thou wast wont to see ;
> Dian's bud o'er Cupid's flower
> Hath such force and blessèd power.—(iv. 1.)

What plant is intended does not appear. We have no
right, as said before, to assign to Shakspere anything of
our own. In a case like the present, when it is plain that
he must have had some definite image before his mind
it is quite permissible, nevertheless, to conjecture. The
allusion may be to wormwood, the famous Artemisia of
the olden time, of which the name "Dian's herb" would
be as nearly as possible a translation.. Not impossibly
he was thinking of the *Agnus castus*, famous in the
middle ages as "a singular medicine for such as would

live chaste," thus fittingly dedicated to the crescent-crested goddess of the virgins,

Queen and huntress chaste and fair,

of Ben Jonson's beautiful hymn. The *Agnus castus* is a pretty little shrub, some five or six feet high, indigenous to Greece, Syria, and Asia Minor, where it grows upon the banks of rivers, and whence it was first brought to England in 1570. The shoots are long and pliant; the rather large leaves are opposite and digitate; the leaflets narrow lanceolate. The flowers, small and white, or with a purplish tinge, grow in loose terminal clusters, and are produced in September.

THE PRIMROSE.

The primrose, like the violet and the pansy—a truly aboriginal British plant—is said to be more plentiful in our own island than in any other part of Europe of equal extent. Upon the Continent it is less widely diffused than either of the above-named; the southern localities lie chiefly among the mountains, so that by those who travel towards the Mediterranean the pleasant sight of it is gradually lost, whence also, very plainly, the infrequent notice of this beautiful wild-flower by authors of earlier date than the fifteenth century. Even the name, in the original shape, is not by birthright its very own, the germ being found in the old Italian appellation of the common field daisy, which in the time of Dante, Petrarch, and

Boccaccio, was *fiore di prima vera,* "the flower of the early spring." Subsequently this name was passed on to the cowslip, which retained it, abreast of the daisy, till as late as the time of Matthiolus, in whose curious old *Epitome,* pp. 653, 655, and 883, there are capital wood-cut representations both of the daisy and the cowslip, though the primrose is not even mentioned. *Fiore di prima vera* became, by contraction, *primaverola.* This, moving northwards, was changed, by the French, into *primeverole.* Then, passing into England, and to the flower before us, it became *primerolle,* as in Chaucer, in the *Miller's Tale,* 82, and in Lydgate, who began to write before the death of Chaucer, producing innumerable little pieces, many of which are fashioned after the manner of his great predecessor, but who again slightly changed the spelling—

> The honysoucle, the froisshe prymerollys,
> Ther levys splaye at Phebus up-rysing.*

From primerolles the transition to "primrose" was quite easy, though there is no evidence as to when it took place. For a long time, moreover, the application of the name was indefinite, one of the plants originally called primrose being the common privet. It was Shakspere who, as in the case of the sweet violet, gave the name its final anchorage. When *he* talks of the primrose we know exactly what is meant; and never since his time has the

---

* In that charming little poem, "Lydgate's Testament."—Percy Society's "Early English Poetry," vol. ii., p. 242.

name borne, in its pure and simple form, any other
signification than the cherished one. Can we be otherwise than grateful once again ?

Including the line in *Venus and Adonis*,

> Witness this primrose-bank whereon I lie ;

including also the pretty reference in the *Midsummer
Night's Dream*, i., 1,

> In the wood where often you and I
>   Upon faint primrose-beds were wont to lie,

mention of it is made by Shakspere upon eight distinct
occasions. The characteristics which naturally most
impressed him, being those upon which a poet would
first alight, were its very early development, and the
peculiarly soft and delicate hue of the petals, so different
from that of the pilewort, or "lesser celandine," *Ficaria
verna*, the constant companion of the primrose, and
which shines, from first to last, with golden lustre.*
Hence the favourite Shaksperean epithet of "pale":—

> I would be blind with weeping, sick with groans,
> Look pale as primrose, with blood-drinking sighs,
> And all to have the noble duke alive.
> *2nd Henry the Sixth*, iii., 2.

---

* The pilewort, if we press for minute accuracy in the association, is much more emphatically the herald of spring than either the
daisy or the cowslip. The former blooms more or less all the year
round, and the latter often in the autumn, whereas the pilewort
never opens until that sweet season when the trees, though there
are leaves upon them, give no shade.

Again in that most lovely passage where Imogen (Fidèle),
lying in a trance, is believed by Arviragus to be dead:—

> With fairest flowers,
> Whilst summer lasts, and I live here, Fidèle,
> I'll sweeten thy sad grave. Thou shalt not lack
> The flower that's like thy face, pale primrose, nor
> The azur'd harebell, like thy veins; no, nor
> The leaf of eglantine, whom not to slander,
> Outsweeten'd not thy breath.—*Cymbeline*, iv., 2.

And once more, in the supreme lines in the *Winter's
Tale*—we know them already—which, though quoted a
thousand times, are for ever new and fragrant. Perdita,
receiving her father's guests at the rustic festival, bids
them welcome, looking first at the elders, and with
courtesy so respectful and refined, so amiable withal, so
unhesitating, and yet so maidenly, that none but
Shakspere could have depicted it:—

> Give me those flowers there, Dorcas—Reverend Sirs,
> For you there's rosemary and rue: these keep
> Seeming and savour all the winter long.
> Grace and remembrance be to you both,
> And welcome to our shearing!

Then comes the modest talk with Polixenes anent the
carnations: then, turning to Florizel, yearning for spring
flowers, the incomparable outpour,

> O Proserpina,
> For the flowers now that frighted, thou lett'st fall
> From Dis's wagon! Daffodils,
> That come before the swallow dares, and take
> The winds of March with beauty.
>
> .   .   .
>
> Pale primroses,
> That die unmarried, ere they can behold

> Bright Phœbus in his strength, a malady
> Most incident to maids; bold oxlips, and
> The crown-imperial; lilies of all kinds,
> The flower-de-luce being one! O these I lack
> To make you garlands of; and my sweet friend,
> To strew him o'er and o'er.
>   FLORIZEL.—What! Like a corse?
>   PERDITA.—No, like a bank for love to lie and play on.

Perdita, child of fortune, has for her distinguishing trait, that most charming of all the womanly virtues, love of truthfulness. Full of all sweet and tender emotions, confiding, and in her own warm heart incomparably faithful, her zeal in friendship is never chequered by a thought that it can be mistaken; without any pretension to power, void of all love for vehement outbursts, she is always, as truthful women always are, natural, graceful, and easy; whoever else may be insincere, it is hers to be frank as the sunshine. Hence, without a pause, the beautiful rejoinder, continuing :—

> *Not* like a corse; or if—not to be buried,
> But quick, and in mine arms. Come, take your flowers.
> Methinks I play as I have seen them do
> In Whitsun pastorals: sure this robe of mine
> Doth change my disposition.

Florizel may well make his enraptured reply :—

> What you do,
> Still betters what is done. When you speak, sweet,
> I'd have you do it ever: when you sing,
> I'd have you buy and sell so; so give alms:
> Pray so; and, for the ordering your affairs,
> To sing them too. When you do dance, I wish you
> A wave o' the sea, that you might ever do

Nothing but that; move still, still so, and own
No other function. Each your doing,
So singular in each particular,
Crowns what you are doing in the present deeds,
That all your acts are queens.

Perdita is the most accomplished of Shakspere's women. To her unassailable purity, her infinite sensibility, she adjoins aptitude for all sprightly feminine pastimes and recreations; though she may not play chess, like Miranda, she can sing, she can dance, she is pious, alive to the inner beauty of wild nature, rejoicing in the spectacle of the daffodils and the swallows, and of boundless resource in regard to the mental acquirements that bespeak culture. Hence she stands full front as the very type and model of perfect womanliness, which consists not alone in being loveable, but perennially companionable. Appearing upon the scene so late, and remaining with us for so short a time, it is somewhat remarkable that her delicious character should be so perfectly delineated.

The primroses she wishes for die, she says, "unmarried." Had the words been spoken to-day, the idea might have seemed Linnæan. Not so is it with Perdita. Phœbus is the sun. To the flowers he is as a bridegroom. They wait his arrival; their wedding-day is when his beams fall bright and warm into their tinted cups:—the primroses, they come all too soon; like maidens who die in their teens, they just glance at him, then fade away.

Two of the allusions are curiously metaphorical. That so pleasing a flower should be used as an image of personal comeliness, as often happens in the literature

of the middle ages, is not surprising.　Chaucer thus uses
it in the *Miller's Tale* above quoted; Spenser also in
Colin Clout:—

> A fairer nymph yet never saw mine eye,
> She is the pride and primrose of the rest.

Roger Ascham, in 1570, applies it to well-favoured youths,
those "two noble primroses of nobilitie, the young duke
of Suffolke and Lord Henry Matravers."* Dr. Prior
quotes from an old writer, in reference to Thomas à
Becket, that he "would sumtyme for his pleasure make a
journey of pylgrymage to the prymerose peerlesse of
Stafforde." Who the lady was the historian does not
say. The pretty phrase he thought her worthy of became
the appellation of the wild English narcissus, the *Narcissus biflorus* of the botanists, which bears it to this day
in country gardens. Somewhat singular is it, then, that
Shakspere, in his metaphorical use, employs the word
not thus, but as a preacher might:—

> But good my brother,
> Do not, as some ungracious pastors do,
> Show me the steep and thorny way to heaven,
> Whilst like a puff'd and reckless libertine
> Himself the primrose-path of dalliance treads,
> And recks not his own rede.—*Hamlet*, i. 3.

So again in the porter's exclamation in *Macbeth*, "I had
thought to have let in some of the professions that go
the primrose-way to everlasting fire" (ii. 3). In each
instance the words are virtually a repetition of what is
said by the honest old clown in *All's Well that Ends*

---

* The Scholemaster, ed. 1571, p. 241.

Well, iv., 5—"I am for the house with the narrow gate, which I take to be too little for pomp to enter. Some, that humble themselves, may; but the many will be too chill and tender; and they'll be for the flowery way, that leads to the broad gate and the great fire." "Recks not his own rede," in the *Hamlet* verse, means, it hardly needs the pointing out, "does not abide by or act up to his own doctrine."

The three passages, taken together, may be accepted as a faithful disclosure of Shakspere's own private and personal views of the essence of Christianity. No man ever stood up more energetically than he, for the sum and substance of scripture teaching, which is that men are saved, or the reverse, not according to what they think, but according to what they do,—"Their works do follow them." Shakspere, though we have no clue to what in these days would be called his "denomination," was a soundly religious man. That in his prose we meet at times with indelicacy and grossness of speech that stands in very painful contrast with his accustomed chaste nobleness, is unhappily quite true; but this, as said before, belongs to the colloquial coarseness of the age in which he lived, and not to himself. In Shakspere there is never any deliberate and flippant irreverence; there is none of the artfully disguised profanity, none of the vicious suggestion, and sapping of good morals, which in the modern novel is often so harmful. A man cannot be void of religion when his writings teem with thoughtful and affectionate reverence of Christian well-

doing and of womanly worth, and who never fails in
gentle enforcement of the divine beauty of peace and
mercy, justice and an upright life. Very interesting is
it to observe, also, how very slight is the reflex in
Shakspere of any particular theological system. Dante
is full of the deep colouring of the theology of the age
in which he lived; and among Shakspere's successors,
not one is more conspicuous in this respect than the
author of *Paradise Lost*. In the main, both Dante and
Milton were no doubt in advance of and superior to the
current theology of their time, but neither of them was
wholly free from the influence of doctrines which in
the grasp they took of inferior minds often became
terrible, but which in Shakspere find little expression.
Shakspere's large perception of the first principles of
Christianity is disclosed, we may here remember, in a
very striking manner, in the history of the experience of
Macbeth—"If human nature," it has been well observed,
"even in its worst hours, had merely to struggle with
itself, the problem of responsibility would be far easier
of solution." Shakspere, in *Macbeth*, illustrates the more
fearful unseen element in human temptation, under the
figure of the three witches. The place they hold in
the drama is not given them for simple theatrical effect.
It is needful to the consistency of the life and crime of
the principal actor. Shakspere's intimate acquaintance
with Holy Scripture has many times been illustrated.
Dr. Wordsworth (Bishop of St. Andrews) pronounces
the simple truth:—"Take the entire range of English

literature; put together our best authors who have written upon subjects not professedly religious or theological, and we shall not find, I believe, in them all united, so much evidence of the Bible having been read and used as we have found in Shakspere alone."

The one remaining allusion to the flower under notice is the other in *Cymbeline*, where the Queen, who has previously commanded—

> While the dew's on the ground, gather those flowers;
> Make haste, who has the note of them?

now says:—

> So, so, well done, well done:
> The violets, cowslips, and the primroses
> Bear to my closet.

THE COWSLIP.

The cowslip, diffused in England quite as widely as the primrose, though usually preferring to dwell in the open meadows, where the flowers can mingle sweetly with the rising grass, is in the south of Europe very conspicuous. Hence, as regards the primulas in general, it holds the place of honour with the pre-Shaksperean draughtsmen. Old Brunfels (1532) figures it under the name of *Herba Paralysis.* In Fuchsius (1542) it is represented as *Verbasculum odoratum;* in Lobel (1576) as *Primula pratensis.* To Shakspere the flower must have been very dear, familiar as the violet, and no flower would he pluck more frequently, a circumstance that

makes one wonder so much the more that over the
cowslip he should have made the one solitary slip in
exactitude of description, which shows in contrast so
striking to his otherwise strict accuracy.   We have dealt
with the matter already; it will be enough, accordingly,
for completeness' sake, to quote the passage anew:—

> On her left breast,
> A mole, cinque-spotted, like the crimson drops
> I' the bottom of a cowslip,

and to repeat that wise people, when in the presence of
nature and of supreme art, never allow themselves to
waste their time in more than a glance at errors and
blemishes, seeing that the longest life is still too short for
grasp even of a tithe of the true and glorious.   Before
you begin to look for flaws and defects, much more
before you begin to censure, assuming to be an arbiter
before you have learned how to be a student, be sure
that you appreciate all the worth.   It is to be remembered also, in regard to art, that the exercise of it by a
master by no means forbids an ennobled representation
of the thing it seeks to pourtray, as distinguished from a
mechanically precise copy.   The profoundest and most
enduring emotions of pleasure are not by any means only
those which arise upon the contemplation of absolute
fidelity on the part of the artist.   It is well to learn how
to be fair enough to allow him, at times, to sacrifice
some little of accuracy in order that he may touch our
hearts with larger imaginings of beauty.   We are not
deceived by what he does, any more than by the sweet

legend of the Graces. Shakspere never misled any one
by saying that the dots in the cowslip flower are
"crimson."

The allusions he makes to the cowslip amount
altogether to six, though found in only four separate
plays—*Cymbeline,* as already indicated, i., 6, and ii., 2;
the *Tempest,* in Ariel's little song,

> Where the bee sucks, there suck I,
> In a cowslip's bell I lie (v., 1);

the *Midsummer Night's Dream,* in two places; and
King Henry the Fifth, once. The Henry-the-Fifth
mention occurs in the very beautiful, though mournful
picture of the ruin and desolation induced by the
calamities of war. A royal and brilliant company is
gathered together. The kings of the rival countries,
Gloster, Warwick, and Burgundy, all are present; the
salutations proper to the time and splendid scene are
duly exchanged; then comes the duke's appeal on behalf
of Peace, the most pathetic of its kind known to literature:—

> My duty to you both, an equal love,
> Great kings of France and England!

Extending to forty-five lines, it is too long to quote just
for the sake of the elegant little phrase in the heart of it,
"the freckled cowslip." Like the "lovely song of one
who playeth well upon an instrument," it does not allow
either of division or curtailing; but it is easy to find, and
there need be no distrust as to the certainty of the
reward.

1

The allusions in the *Midsummer Night's Dream* are found, one, facetiously, in the masque (the story of Pyramus and Thisbe burlesqued) which even in rehearsal had already moved the spectators to " merry tears,"

> These lily brows,
> This cherry nose,
> These yellow cowslip cheeks ; —

the other in that charming flow of poetry wherewith the fairy first announces herself :—

> Over hill, over dale,
>   Thorough bush, thorough brier,
> Over park, over pale,
>   Thorough flood, thorough fire,
> I do wander everywhere,
>   Swifter than the moonès sphere,
> And I serve the Fairy queen,
> To dew her orbs upon the green.
> The cowslips tall her pensioners be ;
> In their gold coats spots you see ;
> These be rubies, Fairy favours,
> In those freckles live their savours :
> I must go seek some dewdrops here,
> And hang a pearl in every cowslip's ear.—(ii., 1.)

As in the story of the "love-in-idleness," Shakspere here again most elegantly lures the imagination to thought of his royal mistress Queen Elizabeth :—

> The cowslips tall her *pensioners* be.

To-day we think of pensioners as no more than dependents, usually poor and disabled, upon the special generosity of the rich. The "pensioners" of Elizabeth's court and retinue consisted of a band of fifty of the handsomest and the tallest men, still in the pride of youth, it was possible to select from the best families in the realm. Their dress was of extraordinary splendour, overlaid so heavily with gold embroidery, that it seemed to consist of little else than gold; jewels, Shakspere's "rubies," adding sparkle. So high did these gentlemen stand in the popular regard, that Dame Quickly, in the *Merry Wives of Windsor*, reckoning up all the gay and noble visitors she can think of, concludes with "And there has been earls—nay, which is more—pensioners" (ii. 2), unless, indeed, poor good-natured and generous, though ignorant woman, she is simply dazzled by their splendid apparel, and fancies that dress and gold constitute nobleness. Shakspere, there can be little doubt, had personal knowledge of their appearance, as the pensioners would form part of the royal train at the time of the queen's famous visit to the Earl of Leicester at Kenilworth, in 1575, ten or eleven years before the poet went up to London, where again he would have opportunities of seeing them.

"Tall" carries a twofold meaning. It preserves the idea of the distinguished stature of the pensioners; it is appropriate also to the idea of the tiny creatures who, with Titania for their own particular mistress, could, when alarmed, "creep into acorn cups." Drayton, in the *Nymphidia*, gives to the latter a very pretty echo:—

> At midnight, the appointed hower;
> And for the queene a fitting bower,
> (Quoth he), is that tall cowslip flower,
> On Hipcut hill that groweth.

The five rubies, the "spots," are the gift of the Fairy queen. Properly therefore, the line should be printed "These be rubies, Fairy favours."

The derivation of the name is obscure. All that is certain is its origin with the Anglo-Saxons, with whom it was the cú-slyppe or cú-sloppe. Ben Jonson, unluckily, is quite astray when thinking he has the sense, he speaks of "Bright dayes-eyes and the lippes of Cowes."

THE OXLIP.

The oxlip, scattered, like the cowslip and primrose, all over England, though nowhere an abundant plant, is, botanically, more interesting perhaps than either, since it combines the characters of both. These two plants appear, in truth, to be divergent expressions of a single type, the cowslip being a contracted or concentrated form of primrose, the sulphur-yellow exalted into golden yellow, and the five tawny watery rays of the latter brightened into well-defined orange spots. In the oxlip these characters anastomose; and precisely after the same manner, the geographical predominance is where the two plants most distinctly mingle and overlap, as in old England. The name rests obviously upon a mistaken understanding of Ben Jonson's etymology of "cowslip," meaning the flower which to the cowslip is in dimensions what the ox is to the cow. Fuchsius gives a drawing of it under the name of *Verbascum non odoratum.* Lobel, also, as *Primula pratensis inodora.*

Shakspere mentions it in the celebrated song in the *Midsummer Night's Dream*, "I know a bank," and again in the *Winter's Tale*, iv., 3, where Perdita so excellently distinguishes this flower under the epithet of "bold."

Shakspere was fortunate in not being troubled with botanical uncertainties. "Oxlip" has long since become the name of two or three different things, the precise relationship of which, one to the other, is a problem. The rich old Elizabethan flower—that one which seems to oscillate between cowslip and primrose—is now considered to be an oxlip only by courtesy. Another plant has put in claims to be considered the genuine oxlip; and if the name is admitted to be the rightful property of whichever of the two may seem best entitled to the rank of "species," then the Shaksperean flower must give way. The second kind of oxlip is the plant called in books, *Primula elatior* and *Primula Jacquinii*. Though occurring in several parts of continental Europe, it was discovered to be British only in 1842, when Mr. Doubleday noticed it in one or two of the eastern counties. The flowers of the *elatior*, instead of being erect, or nearly so, as in Perdita's oxlip, are inclined to be pendulous. They are paler in colour, not so flat, and deficient in the five beautiful bosses so conspicuously surrounding the orifice in the centre of the corolla, alike in the Shaksperean plant, and in the cowslip and the primrose. The scent also is different, not being primrose-like, as in Perdita's flower, but somewhat similar to the less inviting odour of the grape-hyacinth. A very marked technical distinction is observable also in the seed-capsules, which in the *elatior* are narrow-oblong, and of the same length as the calyx. Solve the problem as science may, one thing, happily, is changeless, the reality to our hearts of the flower as it is in Shakspere.

Chapter Seventh.

THE DAISY.

> He says he loves my daughter,
> I think so too; for never gazed the moon
> Upon the water, as he'll stand and read,
> As 'twere, my daughter's eyes; and, to be plain,
> I think there is not half a kiss to choose
> Who loves another best.—*Winter's Tale*, iv., 3.

HE daisy — the original *fiore di prima vera* — appears in Shakspere upon five occasions. It is one of the flowers gathered by Ophelia, appearing first when she enters the royal presence distracted, then in the "fantastic garlands" she sought to hang upon the willow "aslant the brook." Next we have it in *Cymbeline*, iv., 2, when Lucius and his soldiers, moved to sympathy by weeping Fidèle, prepare to inter the headless corpse of Cloten:—

> The boy hath taught us manly duties: Let us
> Find out the prettiest daisied plot we can,
> And make him with our pikes and partizans
> A grave.

The quiet beauty of the "pearled Arcturi of the earth"
as they dot the greensward before it shoots, is adverted
to in the *Rape of Lucrece :—*

> Without the bed her other fair hand was
> On the green coverlet, whose perfect white
> Showed like an April daisy on the grass.

Lastly, with the neatest of little epithets, it leads off the
song at the end of *Love's Labour's Lost:—*

> When daisies pied, and violets blue,
> And Lady-smocks all silver-white,
> And cuckoo-buds of yellow hue,
> Do paint the meadows with delight.

"Pied" literally means "painted," or many-coloured, as
when Caliban says of Trinculo, in the *Tempest*, iii., 2, the
usual dress of a jester being motley, "What a pied
ninny's this!" As used also so picturesquely in Sonnet
xcviii. :—

> From you I have been absent in the spring,
> When proud-pied April, dressed in all his trim,
> Had put a spirit of youth in everything—

lines which recall that other charming passage in *Romeo
and Juliet*, the play that teems, more than any other, with
the vivacities of youth :—

> When well apparell'd April on the heel
> Of limping winter treads.—(i., 2.)

Applied to the field daisy, it felicitously describes the
mound of gold, set round with milk-white rays, the latter
very generally tipped with rosy crimson—whence the
"crimson-circled star,"—which gained for this innocent

little plant its earliest known name, Bellis, or the "pretty one." Under this still current appellation—a rare circumstance in botany—the daisy has mention as far back as the time of Pliny, who speaks of it as a flower of the meadows, useful in medicine. The etymology of the English name, as fairly indicated by the spelling, is found in the Anglo-Saxon *dæges-eage*, "'the eye of day." The daisy, says Chaucer,

> That well by reason men may call it may
> The deisie or els the eye of the day.

THE DAFFODIL.

The daffodil—the flower of all others most intimately associated with the name of Perdita—is, like the pilewort, peculiar to early spring. Coming "before the swallow dares," taking "the winds of March with beauty," it counts with the genuine harbingers of the approaching season, and well deserves its pretty old synonym of Lent-lily, or the spring-tide flower, "Lent" signifying, primarily, the time when the severity of winter passes away and the days begin to lengthen and grow milder. The application of the term to a particular ecclesiastical period has no further significance than such as arose upon the contemporaneousness of the religious observance with the cheerful onward and upward spring movement of living nature, declared as it is by nothing more conspicuously than the welcome daffodil. In another part of the same play we find it mentioned for

the same reason, though by a personage very different
from the gentle lady of old Sicilia :—

> When daffodils begin to peer—
> With, heigh! the doxy over the dale;
> Why then comes in the sweet o' the year,
> For the red blood reigns in the winter's pale.

Autolycus, the singer, as of the verses which succeed,
including that charming line—

> The lark her tirra-lirra chants,—

is generously allowed to be poet as well as rogue, for no
man ever knew better than Shakspere that wayward
nature now and then unites contraries; and, as the
illustrator of all things, he was bound to introduce at
least one example. " Peer " gives a poetical picture
which of its kind stands alone in language :—the "sweet
of the year" is the choicest portion, just as in *2nd Henry
the Fourth*, v., 3, we have the "sweet o' the night:"—the
sense of the concluding word is found in heraldry, a
subject in which Shakspere was no poor scholar; the line
as a whole placing before the mind in the liveliest manner
how beautiful the energy of the vernal sap as opposed to
the dormant and colourless juices of frost and icicle.
Though it may be somewhat homely, this cheerful old
flower, the daffodil, has a very striking individuality.
The tall and shapely vase in the centre, so elegantly
waved and plaited, is unique among British plants. So
is the very peculiar pre-Raphaelite look of the plant,
derived from its attitude, when we see it growing upon
the opposite side of a little stream. When in crowds

mong the grass, there is not one, after the scarlet corn-
oppy, that in brilliant effect successfully rivals it.

Scattered not only all over England, but particularly
bundant in south-western continental Europe, and
iffusing the meadows with golden bloom at the time
f year when the mind is drawn most powerfully to the
:upendous theme of the Resurrection, no wonder that
ıe daffodil received its name. For daffodil is only
n altered form of "asphodel," the Homeric name
f the consecrated flower which the ancient Greeks
ssociated with their Elysium, or the future state of the
lessed. The picture, as drawn by Homer, is no doubt
omewhat gloomy. The flower is there, nevertheless,
rowing in "meadows," the antetype, nearly three
ıousand years ago, of the "sweet fields" of the modern
ymn, and in which the departed are represented as
njoying themselves in congenial ways. "After him I
eheld vast Orion, hunting wild beasts, and the swift-
ıoted son of Æacus . . joyful because I had said that
is son (still upon earth) was very illustrious." (*Odyssey*,
i. and xxiv.) What particular flower Homer intended is
ndiscoverable. It may have been the "poets' narcissus,"
'hich grows wild in Greece; or the polyanthus-narcissus,
'hich extends, in a thin vein of distribution, all the way
·om Portugal to Japan; but it could not be the daffodil
:self, since this does not occur in Homer's country.
'here is no proof even that it was a yellow flower,
lthough the conjectural "asphodelus" of the botanists
; one of that colour. Pindar, it is true, with whom

Paradise becomes insular, talks of the "flowers of gold" that pertain to it, but some of them, he says, grow on "resplendent trees," while "the water feeds others," so that, as elsewhere in the famous old Theban, "golden" plainly means no more than supremely beautiful.   It is enough, however, that the bright hue of the daffodil, and the period of its arrival, when "the time of the singing of birds is come, and the voice of the turtle is heard in the land," recommended it as fit inheritor of the name, which, changing first into affodyl and affodile, at last became " daffodil."   The origin of the initial *d* is undetermined.   Prof. Skeat says that it may either be a survivor of the French d'affrodille or a prefix corresponding to the T in Ted, the short for Edward.

To "take the winds of March with beauty" signifies to charm or captivate them, just as in the *Tempest*, v., 1, Alonso says to Prospero,

> I long
> To hear the story of your life, which must
> Take the ear strangely.

The expression was common in the age, as illustrated in Prov., vi., 25, "Lust not after her beauty in thine heart; neither let her take thee with her eyelids."

THE HAREBELL.

"The azur'd harebell, like thy veins," in *Cymbeline*, iv., 2, is the wild English or sylvan hyacinth, *Scilla nutans*, by the elder botanists called *Hyacinthus non-*

scriptus, or the "unlettered," in reference to the romantic
old myth that the original or oriental hyacinth bore upon
its petals the AI, AI, alas! alas! of the lamenting
god. To-day, by unfortunate confusion of names, this
well-known wild-flower is very generally miscalled the
"blue-bell," the latter appellation belonging properly
to the *Campanula rotundifolia,* the blue-bell of Scotland.
Never arriving until July, and lingering till dark November, "a silent spirit of the solitude," the season of bloom
is quite sufficient to distinguish the genuine blue-bell.
For that Shakspere intends the scilla, and not the campanula, is plain from his putting it beside the primrose.
The scilla is also distinguished as the harebell in
contemporary botanical literature. In Warwickshire, as
in many other parts of England, the harebell floods the
open spaces of green woods with the inexpressibly soft
blue which, looking at a bed of these flowers from a little
distance, always makes it seem as if they were bathed in
azure mist. Shakspere doubtless knew, as well as we do,
that they never look lovelier than early on a fine May
morning, when the sun shines *under* the young leaves of
the trees above; thus, between the blue and green, illumi-
nating both. "Azured," his matchless epithet, is one of
the thousand illustrations in Shakspere that, as some
author has well said, "poetry, like science, has its final
precision," and can no more be re-written than the elements
of geometry. "Blue" means many different things, and
invariably requires some explanatory adjunct, as "light,"
or "dark." Azure denotes the colour of the sylvan

hyacinth and nothing besides. "Try as men will, they
have simply to recur to such epithets, and confess that it
has been done, once for all." It is by minute fidelity
to such details, seemingly unimportant, having nothing
to do with the play itself, and which might be omitted
without affecting the sense or action, that the presence of
the great poet is confirmed. Should he fail to inspire
us, we are wooed, and, failing this, we are still regaled.

The resemblance of the colour to that of the veins,
and the epithet, recall the lines in the *Rape of Lucrece:*—

> With more than admiration he admired
> Her azure veins, her alabaster skin,
> Her coral lips, her snow-white dimpled chin.

OPHELIA'S "CROW-FLOWER."

That a little uncertainty attaches to two or three of the
Shaksperean botanical names has already been mentioned.
One of them, "crow-flower," occurs in the imperishable
Hamlet scene above cited in connection with Ophelia
and the willow—a scene so beautiful in its entirety that
no apology is needed for giving the whole of the context
over again. It is wanted, in truth, for a right understanding of the flower-lines:—

> There is a willow grows ascaunt the brook,
> That shows his hoar leaves in the glassy stream.
> There, with fantastic garlands, did she come,
> Of crow-flowers, nettles, daisies, and long-purples,
> That liberal shepherds give a grosser name,
> But our cold maids do dead men's fingers call them.

There, on the pendent boughs her coronet weeds
Clambering to hang, an envious sliver broke;
When down her weedy trophies and herself
Fell in the weeping brook—her clothes spread wide,
And, mermaid-like, awhile they bore her up;
Which time she chanted snatches of old tunes,
As one incapable of her own distress,
Or like a creature native and indued
Unto that element.   But long it could not be,
Till that her garments, heavy with their drink,
Pull'd the poor wretch from her melodious lay
To muddy death.

Gerard gives crow-flower as the equivalent of *Lychnis Flos-ruculi,* and upon his authority the latter is very generally accepted.   But the lychnis does not come into bloom till a month after the nettles and the long-purples have passed away, and the places of its natural growth are quite different, being in low and marshy ground, often among rushes on the borders of ponds; while the nettle is emphatically sylvan, and the long-purples love the company of cowslips, though these also occur sometimes in moist woodlands.   Shakspere was probably thinking of names current in his time very similar to "crow-flower," and no fewer than three of which were synonyms of harebell.   "Hyacinthus," says Turner, the father of English botany, "is also common in England, though it be not of the beste, and it is called crow-toes, crow-fote, and crow-tees" (vol. 2, p. 18, 1562).   Lyte also calls the harebells "crow-toes," adding that "these be not the hyacinths wherein the mourning marks are printed" (p. 206).   The probabilities thus seem to point to the

Scilla nutans, and assuredly a more elegant idea could not be asked for than of a chaplet presenting the three-fold chord of colour—blue, yellow, and red, flecked with the white stars of the daisy. "Crow-flower" is cited by Britten and Holland as a name still employed in the north for the *Scilla nutans*. The "crow-*bell*," it may be added, was, in Shakspere's time, the daffodil.

OPHELIA'S "NETTLE."

Ophelia's "nettle" is that beautiful deep rich yellow-flowered labiate, to-day commonly called the yellow dead-nettle, or yellow archangel, and by the scientific, *Galeobdolon luteum*. Abundant in shady places, such as are haunted by the harebell, and particularly partial to banks of streams that are steep and shaded, Ophelia could not miss it. She would delight in the flowers, for these make little verandahs round the stem, a ring of them to every pair of leaves, thus forming a most exquisite foil to the harebell both in hue and figure. The name, given because of the resemblance of the leaves to those of the original or stinging nettle, and which covered also the *Lamium album*, was in use long before the time of Shakspere. Turner describes the Lamium as "a kynde of nettel . . . havyng leaves that byte not." Lyte also says "there be two kinds of nettels." The stinging nettle is scarcely developed when the long-purples are in bloom, and the insignificant flowers are not produced before July.

THE LONG-PURPLES.

Long-purples are, without question, the flowers of the crimson meadow orchis, *Orchis mascula*, the beautiful plant so well distinguished from everything else growing wild in England by its hyacinthiné general figure, spotted leaves, and richly-dyed petals. The circumstances of its time and place are precisely such as the context requires, and Shakspere himself contributes a very interesting conclusive proof of the identity, not only in referring to the "grosser name," preserved in the old Herbals, and which in his manly delicacy he withholds from the lips of the queen, but in citing the yet other name, "dead men's fingers." This last was given originally to the *Orchis maculata* and the *Orchis conopsea*, because of the strange similitude found in the palmate and cadaverous tubers, then passed on to the *mascula* as the orchis *par excellence*. Slightly varied, it occurs in the pathetic old ballad that tells of the troubles of "The Deceased Maiden Lover":—

> Then round the meddowes did she walke,
> Catching each flower by the stalke,
> Such as within the meddowes grew,
> As dead man's thumb and harebell blew;
> And as she pluckt them, still cried she,
> Alas ! there's none e'er loved like me.

Certain modern writers have tried to prove that by long-purples Shakspere intends either the *Arum maculatum* or the *Lythrum Salicaria.* That the *Orchis mascula* is

K

meant was shown as far back as 1777 by that admirable old Shaksperean, Dr. Lightfoot, in the *Flora Scotica,* vol. i., p. 515.

"Liberal," in this pathetic Ophelia scene, is employed in the primitive sense of unrestrained, thence licentious and rude, as in *Much Ado,* iv. 1, "a liberal villain;" in *Othello,* i., 2, "a most profane and liberal counsellor," "counsellor" here meaning "adviser;" and in that excellent passage in the *Merchant of Venice,* where we are cautioned as to language in the presence of strangers:

> Gratiano,
> Thou art too wild, too rude, and bold of voice;
> Parts that become thee happily enough,
> And in such eyes as ours appear no faults;
> But where thou art not known, why there they show
> Something too liberal.—(ii., 2.)

In *Othello,* again, it is applied in the sense of "rude" to the north wind:—

> No, I will speak as liberal as the north.—(v., 2.)

"Cold," the quality of snow and of sculptors' marble, means, after the same manner, modest, decorous, incorrupt in speech, the idea which Shakspere invariably connects with genuine womanliness, for Celia, when she once forgets herself, may be considered to be the victim of her boy's clothes. The expression occurs also in the *Tempest,* iv., 1, where the flowers of April serve to make

> Cold nymphs chaste crowns.

Before leaving this celebrated passage, it may be well to

point out the error in some editions, in the printing of
the third line,

Therewith fantastic garlands did she come,

and then, in order to preserve consistency, wantonly
changing "come" into "make." "Therewith" implies
that the garlands were manufactured from the willow,
which is absurd. The sense of the "there" is obviously
thither, to that place, the foot of the tree.

THE LADY-SMOCK.

Shakspere's reference to this delicate and thrice-familiar
field-flower—the only one in which he indulges himself—
was commented upon with explanation, in our opening
chapter. Here we have only to speak of the meaning
of the name, which like almost all others beginning with
Lady, dates from the early middle ages, and is intended
to imply some sort of association with the idea of Notre
Dame, the Virgin Mary, her appearance, her attire, or
her virtues, and occasionally, perhaps, some kind of consecration. Lady's tresses, Lady's mantle, Lady's fingers,
Lady's signet, are all well known. The beautiful green
and white ribbon-like leaves of the digraphis were in
Shakspere's time, Lady-laces; the primrose is still in
Germany, our Lady's key, being the flower with which
she unlocks the spring. The history of the original use
of some of these names is doubtless to be looked for in
classical mythology, in connection with Venus; others
seem to have been derived from the ancient Scandi-

navian mythology, wherein they relate to the goddess Freyja, the "frau" or mother-queen. In any case the final settlement was upon the idea of the Virgin—upon her to whom belongs "Lady-day," March 25th, in the ecclesiastical calendar, and the Lady-chapel in the Gothic cathedral; and none, accordingly, of these names should be written, as sometimes wrongfully, in the plural, or as if the allusion were to ladies in general. The application of the name of Lady-smock to the *Cardamine pratensis,* the flower intended in Shakspere, seems to have been suggested by the peculiarly soft and translucent character of the petals, "silver-white," with faint lilac tinge, and network scarcely visible, all in harmony with ideas of the inmost clothing of purity, such as the mind naturally associates with thought of the Virgin, and which are heightened by the sweet spectacle when these pretty flowers, diffused abundantly, help to "paint the meadows with delight." Lest while contemplating the particular flowers we overlook the singular beauty of the conclusion, it may be well to quote the passage again, and in its entirety :—

> When daisies pied, and violets blue,
> And cuckoo-buds of yellow hue,
> And Lady-smocks all silver-white,
> Do paint the meadows with delight.

For to "paint" does not mean, as in ordinary converse, simply to add colour, though this is perfectly compatible with the general idea, but to enrich or adorn, the use of the word found as far back as 200 B.C., and with precisely

he same application, in the celebrated Roman dramatist,
Plautus. Chaucer also has it in the *Franklin's Tale*,
135:—

> A garden full of leaves and of flowers,
> Which May had painted with her softe showers.

It is in the same sense again that it appears in the
passage in *King John*,—

> To gild refinèd gold, to paint the lily,—

where no reference is intended to such of its race as the
scarlet martagon, Shakspere thinking, probably, when he
wrote the words, of Solomon, who, in "all his glory,'
was still not arrayed like one of these. The four lines,
"When daisies pied," &c., seem to enumerate the flowers
intended by Iris in that charming little interlude in the
Tempest, when she points to the neat finish given by
the industrious farmer to the meadow-borders:—

> Thy banks with pionèd and twillèd brims,
> Which spongy April at thy hest betrims.

Some would have the words to be "peonied and lilied."
But peonies do not arrive until Whitsuntide, and then
only in gardens; and were the ground already dressed
with flowers so splendid, it certainly would need no
betrimming. "Pioned" and "twilled" are old words
denoting use of the spade and mattock.

CUCKOO-FLOWERS.

"Cuckoo-flower," with the old herbalists, was a synonym
of "Lady-smock." The former name covered, however, many other plants, and to fix what it denotes in

particular instances is often impossible.　Such is the case
with the solitary Shaksperean mention, which occurs in
Cordelia's 'description of distracted Lear:—

> Alack, 'tis he: why, he was met even now,
> As mad as the vex'd sea; singing aloud;
> Crown'd with rank fumiter and furrow weeds,
> With harlocks, hemlock, nettles, cuckoo-flowers,
> Darnel, and all the idle weeds that grow
> In our sustaining corn.— (iv., 4.)

None of the plants ordinarily called "cuckoo-flowers"
are associates of cornfield weeds, nor do any of them
bloom at the same period.　All are spring or early
summer flowers.　Very possibly Shakspere may here not
have intended anything definite; for we must not shut
our eyes to the unquestionable fact that the poets universally, at times, introduce names and words for the sake
purely of rhyme, or euphony, or to balance the metre.

There are some similar old wild-flower names, however,
with "cuckoo" for their first portion, which, provincially,
but rather inconsistently, denote plants of late summer
and autumn, and it is just possible that Shakspere may
have had one or other of these in view.　No plant has
ever been more closely identified with the "furrows" of
August than the beautiful sky-blue corn-flower or blue-
bottle, *Centaurea Cyanus*, and according to Britten and
Holland, this identical plant is in some parts still called
"cuckoo-hood."　One of its earliest middle age names
was *flos frumentorum*.　Another name was "hurt-sickle,"
"because," says Gerard, "it hindereth and annoyeth the

reapers, by dulling and turning the edges of their sicles in reaping of corne," a quality that would well entitle it to the epithet of "idle." The cyanus, though now infrequent, appears to have been very general in the Shaksperean age, for "it groweth," Gerard adds, "among wheat, rie, barley, and other graine."

CUCKOO-BUDS.

Shakspere's "cuckoo-buds," may safely be assumed to be the same as the "buttercups" of to-day, especially the *Ranunculus acris,* usually, after the great *Lingua* of the waterside, the tallest of its race. For there are three quite distinct species:—the *acris,* or common meadow buttercup; the turnip-rooted, *Ranunculus bulbosus;* and the creeping buttercup, *Ranunculus repens.* All three are extremely common, but only the two first-named are apt to give a character to the meads, and although they often grow intermingled, the *acris* is the most abundant in spring and early summer.

In gardens, from time immemorial, there has been a variety with double flowers, the analogue of the *aconitifolius,* familiarly "Fair maids of France." This double-blossomed form appears to be the rightful owner of the very old floricultural name "Bachelors' buttons." So, at least, it would appear from Lyte, who on p. 422 gives it as a second appellation of the double "gold-cup," the wild or single form of which, he adds, is very prone to change to the double state. If any other plant went commonly by the same name in the Elizabethan age, it would be the double form of the wild red campion, *Lychnis sylvestris.* In any case, there can be little doubt that they were the flowers of the same ranunculus, when double, which under the contracted appellation of "buttons" were supposed in the Elizabethan age to have some magical influence upon the fortunes of lovers. A reference to them occurs in the *Merry Wives of Windsor*, iii., 2, "He capers, he dances, he has eyes of youth, he writes verses, he speaks holiday, he smells April and May: he will carry it, he will carry it, 'tis in his buttons, he will carry it." "He speaks holiday" means in good and polished language, just as in *King Henry the Fourth* we have "with many holiday and lady terms." "He smells April and May" means like the flowers of those two months.

Chapter Eighth.

WILD-THYME.

Her voice was ever soft,
Gentle, and low, an excellent thing in woman.
King Lear, v., 3.

 HAT dainty and fragrant little denizen of sunny hedgebanks, the *Thymus Serpyllum*, like the two or three preceding flowers, is mentioned by Shakspere only once. The lines in which the name occurs having been set to music, are better known perhaps to people in general than any others in his writings:—

I know a bank whereon the wild thyme blows.
Midsummer Night's Dream, ii., 2.

How much it was prized in the Elizabethan age is shown by the allusion in Shakspere's great contemporary, Francis Bacon, Lord Verulam, in whose essay "On Gardens," originally published in 1597, we have the following:—"And because the breath of flowers is far sweeter in the air (where it comes and goes, like the warbling of music) than in the hand, therefore nothing is more fit for that delight than to know what be the flowers and plants that do best perfume the air, . . . and those which perfume the air most delightfully, not passed by as the rest, but trodden upon and crushed, are three, burnet, wild-thyme, and water-mint." Shakspere, we may be sure, had often noticed this plant, alike as successor of the oxlip and the violet, and when in autumn it strewed its purple for him over the green slopes of airy hills, swelling into those pretty little knolls and bloomy cushions which show the lightness of the soil beneath, or hanging in rosy curtains from the crevices of jutting crags, after the sweet old fashion which in England is so entirely and purely its own.

THE DOG-ROSE.

Shakspere was not unobservant of the dog-roses of the hedge and wilderness, where the tremulous sprays and arching wreaths, covered with little pink concaves, their young hearts golden, toss themselves out with the careless grace so characteristic of this beautiful wild-flower. He speaks of the plant, or its produce, upon four distinct occasions, though in no instance by the appellation of to-day, employing, in three instances, very curiously, the name of the grub which often occupies the heart of the flower-bud—the "canker" by which the petals are

consumed while still in embryo, and which furnishes him
with so many apt comparisons. Thus,—

> As the most forward bud
> Is eaten by the canker ere it blow,
> Even so by love the young and tender wit
> Is turn'd to folly; blasting in the bud,
> Losing his verdure even in the prime,
> And all the fair effects of future hopes.
>
> *Two Gentlemen of Verona*, i., I.

So again from the lips of unhappy Constance, in *King
John*, iii., 4:—

> But now will canker sorrow eat my bud,
> And chase the native beauty from his cheek,
> And so he'll die.

How it happened that the name of the grub was passed
on to the plant does not clearly appear. Mr. H. T.
Riley, in *Notes and Queries* (First Series, x., 153), says
there was a superstition that scratches inflicted by the
prickles were peculiarly harmful and difficult to heal,
causing, as it were, little cancers. Be this as it may,
canker is the name thrice employed by Shakspere:—

> The rose looks fair, but fairer we it deem,
> For that sweet odour that doth in it live.
> The canker blooms have full as deep a dye
> As the perfumèd tincture of the roses,
> .   .   .   .   but they
> Die to themselves. Sweet roses do not so;
> Of their sweet deaths are sweetest odours made.
>
> *Sonnet*, liv.

Then in the beautiful passage in *Much Ado*, i., 3, " I
had rather be a canker in a hedge than a rose in his

grace," which means, "I would rather live in privacy the simple and honest life of nature than be dependent on the favours of a prince." Lastly, in metaphorical use, in *1st Henry the Fourth*, i., 3, we have—

> To put down Richard, that sweet, lovely rose,
> And plant this thorn, this canker, Bolingbroke.

The remaining allusion is to the fruit, so rich and glowing in late October, when the fall of the leaf gives full view of the innumerable little urns:—

> The oaks bear mast, the briars scarlet hips.
>
> *Timon of Athens*, iv., 3.

"Briar" always signifies, primarily, the plant before us, the common wild rose of the hedges. In two or three Shaksperean lines it denotes the garden or cultivated rose, as will be noticed in due course. Occurring in nearly a dozen places besides, with one exception, it points everywhere else, in part at least, to the *Rosa canina*.

THE EGLANTINE.

In the instance referred to, "briar" denotes that delightful species of Rosa which possesses, in addition to deep-hued flowers, the excellent recommendation of scented foliage, whence the familiar epithet of "sweet." In respect of this union of characters, the sweet-briar or eglantine (the latter name, through curious French descent, from the Latin *aculeus*, a prickle) has no rival among its kindred, and scarcely anywhere in botanical nature. Being an indigenous plant, not rare in hedges and thickets, Shakspere quite probably knew it as such, though more commonly as an inmate of the garden, where it always had a place. The old poets introduce it constantly in their pictures of garden pleasures.* Chaucer gives the name of eglantine to one of his ladies. Twice in Shakspere himself, it comes in under the latter appellation, helping, in the *Midsummer Night's Dream*, to "over-canopy" the bank "whereon the wild thyme blows;" and serving in *Cymbeline* for the beautiful comparison already quoted, when with Cytherea and the violet, here connected still more elegantly with Imogen—

> Thou shalt not lack
> The flower that's like thy face, pale primrose, nor
> The azur'd harebell, like thy veins; no, nor
> The leaf of eglantine, whom not to slander,
> Out-sweetened not thy breath.

The simpler name occurs in *All's Well that Ends Well*, that wonderful play which, if less abounding in sweet and grand poetic imagery and description than the other

---

* Thus—
> I would make cabinets for thee, my love—
> Sweet-smelling arbours made of eglantine.
> > *Barnfield.* The Affectionate Shepherd.

> Art, striving to compare
> With Nature, did an arbour green dispread,
> Framed of wanton Ivie, flow'ring faire,
> Through which the fragrant Eglantine did spread
> His prickling arms, entrayl'd with Roses red,
> Which dainty odours round about them threw;
> And all within with flowers was garnished,
> That, when mild Zephyrus amongst them blew,
> Did breathe out bounteous smells, and painted colours shew.
> > *Spenser.* The Bower of Bliss.

principal comedies, supplies one of the most striking proofs given by any of the gay and cheerful ones, of Shakspere's incomparable power of portraiture, laying bare, as it does, the character of Helena:—

> The time will bring on summer,
> When briars shall have leaves as well as thorns,
> And be as sweet as sharp.—(iv., 4.)

Coleridge calls Helena Shakspere's "loveliest character." Mrs. Jameson, who is always right, says in regard to her, "There never was a more beautiful picture of a woman's love cherished in secret, not self-consuming in silent languishment, not pining in thought—not passive and 'desponding over its idol,'—but patient and hopeful, strong in its own intensity, and sustained by its own fond faith. . . . The faith of her affection, combining with the natural energy of her character, believing all things possible, makes them so. It would say to the mountain of pride which stands between her and her hopes, 'Be thou removed!' and it is removed." This is not all. Helena's love is of the kind that shrinks from no exertion, retires before no repulse, resents no unkindness, is incapable of dismay, lives, moves, and has its being in the quenchless conviction that some day, let it only be faithful, the reward shall come, and the heart bathe itself in gladness. How exquisitely consistent, then, that Helena, rather than any one else, should adduce this charming eglantine image, impossible to quote too often for one's own personal sustentation of spirit. Though wintry to-day, never mind: hope on, hope ever:

> The time will bring on summer,
> When briars shall have leaves as well as thorns,
> And be as sweet as sharp.

It is right that in the end she should triumph, and justify the original second title of this inspiring story, "Love's Labour Won."

THE WOODBINE.

By authors of earlier date than Shakspere the name of woodbine was applied to various weak-stemmed climbers, the sense being the encircler of trees, and very specially to the wild clematis, or "Travellers' Joy." Turner restricts it to the sprightly plant which, by common consent, it has now for three centuries denoted in the vernacular, saying, "periclymenon (the ancient Greek appellation, found in Dioscorides) is named . . . in English woodbynde, and in some places of England honysuckle." To Shakspere the double appellation was familiar. In two places the names stand side by side; in one instance only does woodbine occur alone; honeysuckle never does so. The former enriches the picture of the bank, "whereon the wild thyme blows;" this, in addition to its garniture of eglantine, being

> Quite over-canopied with lush woodbine,

while for the coupled names we turn to the *Midsummer Night's Dream* and to *Much Ado:*—

> TITANIA: Sleep thou, and I will wind thee in my arms,
> So doth the woodbine—the sweet honeysuckle—

Gently entwist—the female ivy so
Enrings the barky fingers of the elm.
Midsummer Night's Dream, iv., 1.

HERO· Good Margaret, run thee in the parlour,
There shalt thou find my cousin Beatrice
Proposing with the Prince and Claudio.
Whisper her ear, and tell her I and Ursula
Walk in the orchard, and our whole discourse
Is all of her. Say that thou overheard'st us;
And bid her steal into the pleachèd bower,
Where honeysuckles, ripened by the sun,
Forbid the sun to enter, like favourites
Made proud by princes, that advance their pride
Against that power that bred it; there will she hide her
To listen to our purpose.—*Much Ado*, iii., 1.

Margaret plays her part well; Beatrice is enticed into the garden; she thinks she is unobserved:—Hero resumes with Ursula—

Now begin,
For look where Beatrice, like a lapwing, runs
Close by the ground, to hear our conference;

and then we listen once more to the beautiful imagery so abounding in a play not more remarkable for its brilliant wit and diverting humour than for the poetry:—

URSULA: The pleasant'st angling is to see the fish
Cut with her golden oars the silver stream,
And greedily devour the treach'rous bait;
So angle we for Beatrice, who even now
Is couchèd in the woodbine coverture.

Were it for no more than that the mention of the woodbine brings us into the atmosphere breathed by Beatrice, the occurrence of the name would be matter for rejoicing, for the essential interest of *Much Ado* comes, as in

several of the other plays, of the admirable conjunction
of two separate themes. First, we have a dramatic life-
history; then a profound chapter in philosophical
psychology; yet the two are so blended as not to admit,
even for a moment, of severance. The dramatic evolu-
tion of the tale is perfect. Interwoven with it we have
the fine tissue of personal character which renders the
study of Beatrice inexhaustible; given at the same time
without any lengthy detail of description, therefore with
guarantee of the noblest intellectual pleasure, since this,
as in all consummate works of art, arises upon quiet
sense of brief and minute creative touches, which are
so subtle too that they never seem to be art at all. *Ars
celare artem.* "Beatrice," as Mr. Furnival says, "is the
sauciest, most piquant, sparkling, madcap girl that Shakspere ever drew; and yet a loving, deep-natured true
woman, too." . . . Evidence, so-called! Suspicion!
What are these to her? She knows her friend's pure
heart, where no base thought ever has lodged:—

> O, on my soul, my cousin is belied!

Then when she is married, "we all know what it
means—the brightest, sunniest married life, comfort in
sorrow, doubling of joy." Hazlitt remarks somewhere
that the perfection of woman's nature consists in the
tenacity of her affections, resolving into a combination
of amiable strength with amiable weakness. Beatrice
supplies a perfect illustration. If men do not realise
her character, it is for the same reason, probably, that

L

others do not realise Shakspere in the aggregate—"they have not lived long enough for the reality to come home to them."

But we must not forget the description of the bower into which Hero decoys her. First, it is "pleached," or formed of branches folded together, as in another line in the same play, where Antonio speaks of the "thick-pleached alley" in his "orchard" or garden. The same word is used figuratively in *Antony and Cleopatra*, iv., 12:—

> And see
> Thy master thus with pleach'd arms bending down.

The honeysuckles, in turn, are "ripened by the sun," or so luxuriant in leaf and bloom as to form an impenetrable screen, forbidding "the sun to enter." The phrases are simple enough; they make no pretension; but they illustrate, once again, in the most admirable manner, that the perfection of power is to excel with the slightest possible amount of material.

A word also upon the *Midsummer Night's Dream* passage. Some would have it printed and punctuated so as to imply that the woodbine twists round the honeysuckle:—

> So doth the woodbine the sweet honeysuckle
> Gently entwist.

But let "woodbine" signify what it may, clematis included, weak things seeking support do not twine round other weak things, but round strong ones. In both

:ases, no doubt, in actual nature, two or three of the ong limp stems are occasionally found coiled together, ike the strands of a rope or cable. But this is to be ·egarded only as an accident coming of near proximity; :he exception, and not the rule of life; an occurrence of :he same nature as the reciprocal grip of the tendrils of peas and vetches when there is nothing exterior to take hold of. The original impulse of all twiners and climbers; the impetus which prefigures instinct, just as instinct prefigures reason, is to clasp something substantial and outside, rather than what is another part only of self:—neither woodbine nor clematis ever permanently flourishes except by looking to something stronger. With proper stops, the sense is clear enough. "As the woodbine, otherwise called the honeysuckle, and the female ivy, embrace the elm," Titania means, "so will I, my love, encircle thee." Shakspere's intention in using both names is to make quite plain what plant he refers to. "Woodbine" having been applied to various plants, he is wishful that there shall be no misapprehension, and that the imagination shall be fixed at once upon the Lonicera. One cannot but remember, in passing, that pretty passage in William Bulleyn, the old botanist and preacher, temp. Henry VIII., when, after describing the sweet smell of the honeysuckle after a shower of rain, he goes on to say "it spredeth forth his sweete lilies like ladies' fingers among the thorns."

The common or great hedge bindweed, *Convolvulus sepium*, so distinguished ·for the charm of its white_bell-flowers, so pure and lovely that Virgil employs them as an image of the nymph Galatea, has been thought by some to be the woodbine of the passage before us.   But as we see, there is no occasion to resort to any such conjecture.   The idea seems to have been derived from Ben Jonson, who himself, however, makes a double mistake, since *he* intends the *Solanum Dulcamara.*—

> So, the blue bindweed doth itself enfold
> With honeysuckle; and both these entwine
> Themselves with briony and jessamine,
> To cast a kind and odoriferous shade.
>
> *Vision of Delight.*

Chapter Ninth.

THE GARDEN FLOWERS.

He hath songs for man or woman of all sizes.
Winter's Tale, iv., 4.

EENLY alive to the charms of wild and untrimmed nature, Shakspere could not fail to love a well-furnished and well-kept garden. How he delighted in the contemplation of the little enclosures which Taste, in all ages, has devoted to choice flowers and fruits, is plain from the numberless allusions in the dramas, not simply to gardens, but to gardeners and to horticultural operations, the latter being generally used, moreover, for some beautiful image or comparison, as in the celebrated scene in *King Richard the Second.* The subject of discourse is the forlorn condition of the country induced by the misgovernment of the unfortunate king,

already deposed, the head-gardener giving directions to
his subordinates:—

> Go, bind thou up yon dangling apricocks,
> Which, like unruly children, make their sire
> Stoop with oppression of their prodigal weight.
>
> .   .   .   .
>
> ASSISTANT: Why should we, in the compass of a pale,
> Keep law, and form, and due proportion,
> Showing, as in a model, our firm estate ?
> When our sea-wallèd garden, the whole land,
> Is full of weeds, her fairest flowers choked up,
> Her fruit-trees all unpruned, her hedges ruined,
> Her knots disordered, and her wholesome fruits
> Swarming with caterpillars?
> GARDENER: Hold thy peace:—
> He that hath suffered this disordered spring,
> Hath now himself met with the fall of leaf:
> The weeds that his broad-spreading leaves did shelter,
> That seemed, in eating him, to hold him up,
> Are plucked up, root and all, by Bolingbroke.
> I mean the Earl of Wiltshire, Bushy, Green.
> ASSISTANT: What, are they dead?
> GARDENER: They are, and Bolingbroke
> Hath seized the wasteful king.—Oh! what pity is it
> That he had not so trimmed and dressed his land,
> As we this garden !   We, at time of year
> Do wound the bark, the skin of our fruit trees,
> Lest, being over-proud with sap and blood,
> With too much riches it confound itself.
> Had he done so to great and growing men,
> They might have lived to bear, and he to taste
> Their fruits of duty.   All superfluous branches
> We lop away, that bearing boughs may live:—
> Had he done so, himself had borne the crown,
> Which waste of idle hours hath quite thrown down.

(iii., 4.)

The disordered "knots" are the flower-beds, which in the gardens of the Elizabethan times were marked out with mathematical precision of form and outline, so as to produce in the aggregate, quaint patterns, the entire area of the ground, which was always rectangular, being completely occupied. The idea seems to have been to lay out the garden in some kind of accordance with the domestic architecture of the period, which delighted in straight lines, gables, long terraces, flights of broad stone steps, and in the interior, in long galleries, with bay-windows, and abundance of tracery. It was just such a one as this which is adverted to in *Love's Labour's Lost*, i., i.:—"It standeth north-north-east and by east, from the west corner of thy curious-knotted garden."

In these elaborately-constructed gardens there was never any lawn. The space was wholly devoted to flowers, with a few simple evergreen shrubs, such as box and dwarf yew. Adjoining them were pleasant "alleys" and shaded walks, and bowers and arbours seem to have been an essential feature. The flowers were of the kinds cherished to-day as "old favourites," comprising many brought in from the neighbouring woods and meadows, a good number from continental Europe, and a sprinkling of choice rarities from the Levantine countries, Persia, and northern Africa. Seven-eighths of the contents of the modern English garden, it is almost needless to remark, were quite unknown. Scarcely anything had arrived from America, and not a single species from the Cape of Good Hope, from eastern Asia, or from Australia, then undiscovered. The number of plants in cultivation was still rather considerable, as appears from the old herbalists, especially Gerard, whose massive volume appeared in 1597. So far as it is possible to judge from the allusions in Elizabethan literature, it would seem as if in the Shaksperean age there was more real love for the few flowers then possessed, than is compatible with the *embarras de richesse* of the present day. The enormous quantity now possessed, the incessant arrival of novelties, alike forbid the quiet friendship which three centuries ago there was so little to distract, though there was plenty to sustain, and to invigorate in the sweetest manner.

The catalogue of Shakspere's wild-flowers, as we have seen, is very brief, yet the list of his garden flowers is only half as long, extending to only eight or nine. He had no occasion to mention any more; the great mass of the references fall, as it is, upon only two, the lily and the rose, these having been, from time immemorial, the poets' metaphors for loveliness and purity, especially feminine, and as shown in the feminine cheek, thus, in truth, part of the established vocabulary of civilised man. The *shūshan* and *chābhatstsèleth* of the Hebrews, the λείριον and ῥόδον of the Greeks, the *lilium* and *rosa* of the Romans, are their antetypes; only that while the application of the ancient names is indefinite, and the botanical species intended are often indeterminable, the nearer we draw to the Shaksperean times the more precise they become, Chaucer leading the way in our own country, till,

at last, in Shakspere himself, as in the case of the violet, we rest, and in thankfulness. The actual beginning of the rose and lily usage, like the sources of great rivers, is hidden among the mountains. It is as old, at all events, as the time of Solomon, whose use of the words above cited is only another way of saying " My beloved is white and ruddy," an infinite amount of grand significance lying within. Shakspere reflects the latter when Viola, in *Twelfth Night*, says to Olivia,—

> 'Tis beauty truly blent, whose red and white
> Nature's own sweet and cunning hand laid on.—(ii., 4.)

The comparison comes to the front again in *King John*,—

> Of Nature's gifts thou may'st with lilies boast,
> And with the half-blown rose (iii., 1),—

an allusion so much the more beautiful from the circumstance that the rose is the only flower which is quite as lovely in the opening bud as when full blown. Theocritus so admired the rose when in this condition that he calls the flowers simply calyces.* Shakspere never tires of the twofold citation. See how exquisitely it reappears in the *Rape of Lucrece*, the figure here partly founded upon a later event of history:—

> The silent war of lilies and of roses.

So again in *Coriolanus:*—

> Our veil'd dames
> Commit the war of white and damask in
> Their nicely-gauded cheeks to the wanton spoil
> Of Phœbus' burning kisses (ii., 1);

---

* Idyll, iii., 23.

"Damask" is the colour of the damask or Damascus rose, noted for its fine crimson hue, whence in Pliny the happy epithet, *ardentissima.* This species had also (as to the present day) a white variety, and another that was parti-coloured, which latter is referred to in *As You Like It:*—

> There was a pretty redness in his lip;
> A little riper and more lusty red
> Than that mixed in his cheek.   'Twas just the difference
> Betwixt the constant red and mingled damask.—(iii., 5.)

"Constant" here means uniform or unbroken, as usual in the petals of roses, very few presenting a mixture of colours.   Some think that the other allusion is to the silken fabric called "damask," in which, by skilful crossing of the threads, different shades are produced according to the play of the light.   This last is in any case the intent of the word, most likely, in the famous lines in *Twelfth Night:*—

> She never told her love,
> But let concealment, like a worm i' the bud,
> Feed on her damask cheek.—(ii., 4.)

Apart from the poets' use of rose and lily as names for loveliness and perfection, it is very interesting to observe that these two flowers are representative of the two great sections into which all plants bearing flowers are divided by botanists, namely the exogens and the endogens, or the dicotyledons and the monocotyledons.   Individual species of each section attain greater dimensions, and are of incomparably greater service to man in point of economic utility; but the pentamerous perianth of the

ᴐne, perfectly regular, the petals free, and the calyx and
ᴄorolla quite distinct; and the remarkably symmetrical
ᴀnd sex-partite or tri-partite perianth of the other, the
ᴇlements all congenerate in colour and substance, are
ᴊtill in the front, super-eminently characteristic of the
respective plans of general structure. From rose and
lily we acquire our best ideas of what exogens and
endogens really are; the twofold realm of floral nature
universally acknowledges them the respective queens.
The place and period of the uprise of the two names,
rose and lily, is undetermined. All that can be said with
certainty is that they are of oriental and very ancient birth.

The roses cultivated in English gardens in the Shaksperean age appear to have been the *centifolia*, with its
variety, the *Provincialis;* the above-named *Damascena*,
so called from the glorious old Syrian city, that one
which of all the ancient cities in the world has alone
remained unscathed; the *moschata*, or musk-rose, and
the *alba;* and of these, it would seem, from the very
vague and scanty references to double roses, and the
frequent allusions to the beautiful yellow centre of the
rose in its natural state, the original or single forms alone
were general. The latter, conventionalised, was the rose
of heraldry, as shown by Mr. Boutell, plate xiii., fig. 385.
When it was that the earliest of the roses to arrive
in our country made its appearance no one can tell.
Possibly in the Roman times; more probably in the
early Norman, or during the period of the Crusades, the
enterprises which contributed so greatly to the new birth

of civilisation in Europe, to the diffusion of the love
of art, and of the beautiful in all its forms.   The story of
the Wars of the Roses, beginning A.D. 1455, shows that
the *white* flower was then cultivated as well as the red,
and that it was sufficiently· well known to carry significance not inferior.   We are apt to forget sometimes, in
our attention to prose history, how much Englishmen
owe· to Shakspere in respect of his dealing with those
wars.   Shakspere has made them what the "Tale of
Troy" was to the Greeks, only that the narrator is greater
even than Homer.   As a celebrated statesman lately
told us, "If you take the works of Shakspere from
Richard the Second to *Richard the Third*, inclusive, you
have the Wars of the Roses treated with a vigour and a
variety of conduct, and a multiplicity of incident which
the Iliad cannot excel, combined with a human interest
which no ancient work ever yet commanded."   How
calmly the story opens:—

PLANTAGENET: Let him that is a true-born gentleman,
          And stands upon the honour of his birth,
          If he suppose that I have pleaded truth,
          From off this briar pluck a white rose with me.
SOMERSET: Let him that is no coward, nor no flatterer,
          But dare maintain the party of the truth,
          Pluck a red rose from off this thorn with me.

The nobles take their respective sides; the contention
grows more and more angry as it proceeds; at last come
the words from Warwick, so terribly verified:—

          And here I prophesy: This brawl to-day,
          Grown to this faction, in the Temple Garden,

> Shall send, between the red rose and the white,
> A thousand souls to death and deadly night.
> > *1st King Henry the Sixth*, ii., 4.

Without reckoning the repetitions in this famous scene, which amount to nearly thirty; independently also of the apothecary's "old cakes of roses," dried rose leaves,

> Thinly scattered to make up a show,
> > *Romeo and Juliet*, iv., 1,

and of the mention of rose-water in the *Taming of the Shrew*,

> Let one attend him with a silver basin,
> Full of rose-water, and bestrewed with flowers,

there occur in Shakspere nearly sixty allusions to this flower of incontestable queenliness. Including those already cited, fully one-third of them touch it as an object of nature:—

> The seasons alter; hoary-headed frosts
> Fall in the fresh lap of the crimson rose.
> > *Midsummer Night's Dream*, ii., 2.

> At Christmas I no more desire a rose
> Than wish a snow in May's new-fangled shows,
> But like of each thing that in season grows.
> > *Love's Labour's Lost*, i., 1.

> What's in a name? That which we call a rose,
> By any other name would smell as sweet.
> > *Romeo and Juliet*, ii., 4.

> Cæsario, by the roses of the spring,
> By manhood, honour, truth, and everything,
> I love thee so ——. *Twelfth Night*, ii., 4.

The last-named seems to indicate the time of year of this charming play, so well characterised as "a genuine comedy, a perpetual fount of the gayest and sweetest fancies;" for, laid as the scene is, in the graceful though capricious lady Olivia's own garden, nothing would be more natural than that she should make witnesses of the flowers beside her.

Then we have the long succession of beautiful figurative uses: —

How now, my love, why is your cheek so pale?
How chance the roses that do fade so fast?
Midsummer Night's Dream, i., 1.

Their lips were four red roses on one stalk,
Which in their summer beauty kissed each other.
Richard the Third, iv., 3.

Hail, virgin, if you be, as those cheek-roses
Proclaim you are no less.—*Measure for Measure*, i., 5.

But soft, but see—or rather do not see,
My fair rose wither.—*Richard the Second*, v., 1.

Fair ladies masked are roses in their bud;
Dismasked, their damask sweet commixture shown,
Are angels veiling clouds, or roses blown.
Love's Labour's Lost, v., 2.

Sweet rose, fair flower, untimely pluck'd, soon vaded,
Pluck'd in the bud, and vaded in the spring.
The Passionate Pilgrim.

When Hamlet addresses Ophelia as

O rose of May!

he does not mean the May of the calendar, intending rather the early summer of woman's life, when her cheeks, as well as her hopes, are clad in roseate. Would that all our young ladies would receive the truth that rose-colour is only oxygen in another shape, and that cheeks get it best where the flowers do—out of doors. When, again, Hamlet says,

With two Provincial roses on my razed shoes (iii., 2),

the reference is not at all to the flower *per se.* He means rose-like ornaments made of ribbon, very fashionable in the Elizabethan times, and referred to also by Bottom, in the *Midsummer Night's Dream,* iv., 2, "Get new ribbons to your pumps." Romeo perhaps intends the same when he remarks, "Why then is my pump well-flowered" (ii., 4), though here the reference may be to imitation pinks, or perhaps to its being elegantly pinked or punched with some floral pattern. The musk-rose, the ever welcome species from northern Africa, its long and rambling shoots too weak to stand erect without the assistance of kindly neighbours, and which exhales its delightful musky odour most decidedly in the evening, is associated exclusively with Titania and the fairies.—*(Midsummer Night's Dream,* ii., 2; ii., 3; iv., 1.)

THE LILY.

While the idea of the rose, from the earliest times, has been of something rich and red, covering not only itself, but the oleander and the pomegranate flower—the poetical idea of the lily has been one of whiteness, whiteness of a shining and dazzling kind, or such as would be called lustrous, a whiteness very different from that of milk, and giving the flower a peculiar fitness for

use as an emblem. The ancients, because of this association, gave the word various delightful uses. Homer
puts "lily-voiced" for sweet and pretty sounds, such as
the chirp of the cicala (*Iliad*, xiii., 880). Though,
primævally, a name of very various application, the
lily *par excellence* of the ancients was most probably
that excellent species called by botanists the *Lilium
candidum*, the "common white lily" of the vernacular.
Indigenous to the southern parts of Europe, from Corsica
to Greece and Turkey, it could not fail to be known to
all who noticed flowers. The erect and leafy stems, a
yard high, the noble and well-balanced terminal cluster
of five or six pearly white and fragrant bells, the spotless
petals curving outwards, the swinging golden anthers,
and emerald stigma, would arrest the attention of the
most incurious. This one, at all events, it was which
pious emotion gave to the Virgin, whence the still-current
appellations of Madonna lily and Annunciation lily, not
to mention the thousand beautiful examples met with in
sacred and legendary art. Shakspere's lily was unquestionably the same. He mentions it upon over twenty
distinct occasions, sometimes objectively, more frequently
in metaphor, or with a view to comparison, though not
always in reference to the lovely. Thus,—

> O sweetest, fairest lily!
> My brother wears thee not one half so well
> As when thou grewest thyself.—*Cymbeline*, iv., 2.

> Now by my maiden honour, yet as pure
> As the unsullied lily.—*Love's Labour's Lost*, v., 2.

In *King Henry the Eighth* the lily is used, and in the most beautiful manner, to represent the life-long maidenhood of Queen Elizabeth. She is a babe at the christening font. Cranmer's benedictions remind one of the blessings bestowed by the patriarchs of ancient Palestine, the last, as should be, the flower and crown of all:—

> Many days shall see her,
> And yet no day without a deed to crown it.
> Would I had known no more ! But she must die,
> (She must, the saints must have her), yet a virgin;
> A most unspotted lily shall she pass
> To the ground, and all the world shall mourn her.—(v., 4.)

In *Troilus and Cressida,* iii., 2, "lily" is put for Homer's asphodel above mentioned:—

> O, be thou my Charon,
> And give me swift transportance to those fields,
> Where I may wallow in the lily beds
> Proposed for the deserver.

In an earlier scene of *King Henry the Eighth* there is an incomparably touching allusion to the imagery in the New Testament of the "lilies of the field," described by the divine preacher, first as the produce of "the grass of the field," then as "to morrow cast into the oven :"—

> QUEEN KATHARINE: Shipwreck'd upon a kingdom, where
>   no pity,
> No friends, no hope; no kindred weep for me;
> Almost no grave allow'd me: like the lily,
> That once was mistress of the field, and flourished,
> I'll hang my head, and perish.—(iii., 1.)

The "lilies" of the New Testament, it is well known,

M

are flowers in general, the word having been commonly
thus used in ancient times, or in the broad and general
sense of ἄνθος and *flos.* Moschus, the ancient Sicilian
poet, represents Europa as gathering "fragrant lilies"
in the meads, plainly intending flowers in general.*
"Rose" was often applied in the same way, a very
picturesque example occurring in Pindar.†

On one occasion "lily" is used facetiously, being
introduced in the laughable mock-pathetic burlesque of
Pyramus and Thisbe, performed at the end of the
Midsummer Night's Dream:—

> Asleep, my love?
> What, dead, my dove?
> O, Pyramus, arise,
> Speak, speak.   Quite dumb?
> Dead, dead, a tomb
> Must cover thy sweet eyes.
> These lily brows,
> This cherry nose,
> These yellow cowslip cheeks,
> Are gone, are gone!
> Lovers, make moan!
> His eyes were green as leeks.
> O sisters three,
> Come, come, to me!

Personal beauty being always heightened by darkness of
the eyebrows, "lily brows" well match the cherry nose,
the green eyes, and the cowslip cheeks. A more
ludicrous combination it is impossible to imagine.

---

* Idyll, ii., 32.   † Isth., iii., 36.

THE FLOWER-DE-LUCE.

The flower-de-luce is one of those desired by Perdita:

> Bold oxlips, and
> The crown-imperial; lilies of all kinds,
> The flower-de-luce being one. O these I lack,
> To make you garlands of.

In Chaucer the fleur-de-lis, as we now call it, was the *Lilium candidum*, just described:—

> His nekke was whit as is the flour de lys.

But in the Shaksperean age, and long before, the name was very evidently applied also to various species of Iris—that extensive and charming genus, represented abundantly in Europe, the members of which are distinguished from all other liliaceous plants by their three great outer pendulous petals, familiarly termed the "falls;"—large, erect, and very handsome petaloid stigmas; and the rich and endless diversity of their colours, whence the name, to translate literally, of "rainbow-flower." Many of the exotic species were cultivated in English gardens in Shakspere's time; he would be acquainted also with that strikingly handsome water-side wild-flower, the yellow water-flag, *Iris pseudacorus*, found everywhere upon the continent, and reasonably supposed to have furnished the fleur-de-lis of the French heraldic shield. What particular flower he was thinking of when he wrote "flower-de-luce" there still is no evidence to show. He could scarcely have intended the *Lilium candidum*, since this does not come into bloom till the oxlips are all gone. He knew well enough, however, that the golden one belonged to the insignia of France, applying the name in *King Henry the Fifth*, v., 2, in graceful metaphor, to the French princess—"my fair flower-de-luce;" and using it also upon three occasions, heraldically, in *King Henry the Sixth*:—

> Cropp'd are the flower-de-luces in your arms,
> Of England's coat one half is cast away.—*Part 1st*, i., 1.
>
> I am prepared; here is my keen-edged sword,
> Deck'd with five flower-de-luces on each side.—*Ibid*, i., 2.
>
> A sceptre will I have, have I a soul,
> On which I'll toss the flower-de-luce of France.
> *Part 2nd*, v., 1.

"Of England's coat one half is cast away" refers to the beautiful shield of arms established by Edward the Third for the English monarchy, by combining the three Plantagenet lions with the lilies of France, thereby asserting sovereignty over the latter country, as expressed also in the royal title then assumed, "King of France and England." "One half" of the coat being "cast away" signified, heraldically, loss or forfeiture of dominion in France. Few circumstances in history are more curious of their kind than the retention of the fleur-de-lis in the shield, and thus upon the gold and silver coinage of Great Britain, up to the year 1801, or for two hundred and forty-three years after the loss of Calais. The name is spelt also, in old authors and in early editions, "floure delice" and fleur-de-lys.

THE CROWN IMPERIAL.

Nature produces few plants more remarkable than the crown imperial, *Fritillaria imperialis.* The succulent stem, two feet high, rises out of a tuft of narrow leaves and terminates in a similar tuft, though bare intermediately. Immediately below the latter there is a ring of large tulip-like flowers, of a rich golden yellow, and pendulous. Looking into the flower, as into a bell, at the base of every petal there is seen a white and concave nectary, from which, when the bloom is in perfection, depends a drop of honey. At the first glance it seems to possess six great round eyes carved out of pearl. A native of Persia, it was originally introduced into the royal garden at Vienna about A.D. 1576, and shortly afterwards arrived in England. Shakspere would see it, in all likelihood, in some choice London garden.

THE MARIGOLD.

The marigold, *Calendula officinalis,* mentioned by Shakspere upon five occasions, is, in its most engaging use, still Perdita's:—

> The marigold that goes to bed with the sun,
> And with him rises, weeping.—(iv., 3.)

The flower she intends is the well known ancient inhabitant of the garden—originally from southern Europe,—celebrated alike for its economic service, and for the extreme sensitiveness of the flowers to the solar ray.

The last-named feature was noticed two thousand years ago. It was into the calendula, it would seem, though the description is somewhat loose, that love-sick Clytie was transformed:—

> Illa suum, quamvis radice tenetur,
> Vertitur ad Solem; mutataque servat amorem.
>> (She, though held fast by a root, still turns toward her beloved Sun, and though changed in shape, forgets not to be faithful.—*Met.* iv., 269-270.)

In the calendula, also, we have the original *heliotropion*, in Latin *solsequium*, whence, with lapse of time, the French *souci.* Sunflower and turnsole, though we now apply the former name to an American plant with disk and rays of sunlike splendour, were, in the first instance, it would appear, appellations of the same. The literature of the middle ages abounds with allusions to it. " The marygolde," says Lyte, " hath pleasant, bright, and shining yellow floures, the which do close at the setting downe of the sunne, and do spread and open againe at the sunne rising" (Book II., chap. xiii., p. 163). "Some," continues Lupton, " call it *sponsa solis*, the Spowse of the Sunne, because it sleepes and is awakened with him." In Prime's *Consolations of David*, applied to Queen Elizabeth, in a sermon preached at Oxford, November 17, 1588, we have "David found it true that he should not have been heretofore at any time, and therefore professeth that for the time to come he would be no marigold servant of God, to open with the sun, and shut with the dewe." In writing the lines above quoted from the *Winter's Tale* it is important to insert the comma, often omitted, after "rises," so that the stress may fall upon the "weeping"—Clytie's tears—the dew-drops which give perfection to the picture.

The English appellation of this famous old flower is not, rather singular to say, its own by birthright, nor has it anything to do with the name of "Mary." If not the literal translation, it is the etymological representative and counterpart of the old Greek name of the flower to-day called, by pleonasm, the "marsh-marigold"— *helichryson*, as in Theocritus, ii., 78, literally the golden-flower-of-the-marshes. How it came to be transferred from the *Caltha palustris* to the calendula does not appear.

The other Shaksperean allusions occur in Sonnet xxv.:

> Great princes' favourites their fair leaves spread
> But as the marigolds at the sun's eye.

In the *Rape of Lucrece,*

> Her eyes, like marigolds, had sheathed their light,
> And canopied in darkness sweetly lay,
> Till they might open to adorn the day.

In *Pericles*, iv., 1.,

> The purple violets and marigolds
> Shall, as a chaplet, hang upon thy grave,
> While summer days do last.

And with slight variation of the name, in *Cymbeline*, ii., 3,

> And winking marybuds begin
> To ope their golden eyes.

THE CARNATION AND THE GILLIFLOWER.

These two, yet again, are Perdita's own:—

> The fairest flowers o' the season
> Are our carnations, and streaked gilliflowers,
> Which some call nature's bastards; of that kind
> Our rustic garden's barren, and I care not
> To get slips of them.

Over the carnation there is no doubt. Introduced from the continent, probably in the time of the Normans, it took at once the place of honour in English gardens never since challenged. All the old poets have something to say in its praise, now and then indicating the derivation of the name, which is simply a condensed form of "coronation"—a flower adapted for use in chaplets, and which has never been given to anything else. "Gilliflower," often supposed to be the same, had a much wider meaning. In Lyte, 1576, it covers the carnation, the pink, the sweet-william, the marsh-lychnis, the soapwort, and a couple of silenes, all members of the same botanical family; also various crucifers, including the hesperis, the two Matthiolas, the dentaria, and the wallflower; and, besides these, the Hottonia and the *Tagetes patula.* Essentially, nevertheless, "gilliflower" was understood to mean certain varieties of the *Dianthus Caryophyllus* (the typical form of which is the carnation *ipsissima*), as shown by the derivation supplied by "caryophyllus," the latter word referring to the clove-like

odour again of the carnation *ipsissima.** The varieties
in question were doubtless those which to-day are called
bizarres, and which the florists of the Elizabethan age
thought to multiply by means of the odd and futile
process described in that curious old black-letter volume,
Hyll's *Art of Gardening,* 1574, p. 88. Perdita, whose
character, as we have seen above, is distinguished by its
love of truthfulness, will have none of these artificial
things. She is no horticultural experimentalist: she
takes the flowers just as God made them. In her
innocent ignorance she thinks that all the "pied gillies"
have been obtained by the artificial process she detests.
She admits that as companions of the carnation, they
count with the "fairest flowers o' the season," but to give
them a personal welcome she declines:—

> Of that kind
> Our rustic garden's barren, and I care not
> To get slips of them.
>
> .   .   .   .   .
>
> I'll not put
> The dibble in earth to set one slip of them:
> No more than were I painted, I would wish
> This youth should say 'twere well; and only therefore
> Desire to breed by me.

---

* No botanical name ever experienced so many mutations. In
the literature of the fifteenth and sixteenth centuries may be found
gyroflée, giroflée, gilofer, gillofer, galofer, gylofre, girofle, gelofer,
gyllofer, gelouer, gillyvor, gelyfloure, gilyfloure, gilofloure, gillo-
floure, gelliflower, gilloflower, no fewer than seventeen different
spellings, and probably others not observed.

How modestly she disclaims personal acquaintance with
the process:—

POLIXENES: Wherefore, gentle maiden,
               Do you neglect them?
PERDITA: For I have heard it said
               There is an art which, in their piedness, shares
               With great creating nature.

Polixenes—Shakspere—knew nothing of "cross fertili-
sation," the practice of which has given to modern
horticulture a complexion so new and grand—though in
a measure foreseen and here foretold by him, as well as
by Bacon. So much the more remarkable, accordingly,
becomes the reply:—

          Say, there be;
      Yet nature is made better by no mean,
      But nature makes that mean; so, o'er that art
      Which, you say, adds to nature, is an art
      That nature makes.

   .   •   •   •   •

           This is an art
      Which does mend nature—change it rather, but
      The art itself is nature.

The discourse between these two, Perdita and Polixenes,
illustrates over again that other admirable characteristic
of Shakspere—the disclosure that he knows everything,
without ever advertising his knowledge. The "for,"
before "I have heard it said," means "because."

Another name, in the middle ages, for the carnation,
was "sops-in-wine," or simply "sops," given because
the flowers were supposed to impart flavour, especially

the sweet wine presented to brides after the wedding
remony. Hence, in the *Taming of the Shrew*, iii., 2 :—

> Quaff'd off the muscadel,
> And threw the sops all in the sexton's face.

A smaller kind of Dianthus, the pink, is generally
pposed to be referred to in *Romeo and Juliet*, ii., 4,
1en Mercutio figuratively styles himself "the very pink
courtesy," or courtesy in perfection. In the follow-
g scene Nurse says that Romeo is *not* the flower of
urtesy, here, perhaps, playing upon the name of the
autiful crimson amaranth to-day called the prince's
ather, in Shakspere's time the "floure-gentle."

"Coronation," literally "garland-flower," the original ✓
rm, as above said, of "carnation," serves as the key-note
some other very interesting words. The diminutive
the Latin *corona*, upon which it rests, is *corolla*,
erally "a little crown," very elegantly applied, in
tanical language, to the aggregate of the petals of a
wer, which, although often insignificant and without
mmetry, in very many cases—those, no doubt, which
ggested the application of the name—may quite fairly
said to be disposed chaplet-wise around the interior
rtions. The employment, by the ancients, of crowns
d chaplets wrought of leaves and flowers, especially for
stowal in testimony of honour and approval, was very
tensive, and has innumerable illustrations in classical
erature. Chaplets thus given were things ensuing
aturally upon what had preceded, as honours must

needs, especially when symbolical. *Corollarium*, "a little garland," thence came to denote, frequently, a natural result or sequence of any other kind; and thence something supplementary or additional. When, accordingly, in the *Tempest*, iv., 1, Prospero says,

> Now come, my Ariel, bring a corollary,
> Rather than want a spirit,

he means rather than be deficient, bring a surplus.

THE COLUMBINE.

The columbine, *Aquilegia vulgaris*, though a plant wild in England, was probably, like the sweet-briar, best known to Shakspere as a cultivated garden-flower. It appears, slightly, with metaphorical purpose, in the pageant which ends *Love's Labour's Lost;* and again in *Hamlet*, here in reference to some emblematic or superstitious idea the traces of which are obscured. "There's fennel for you and columbines." Ophelia employs it, seemingly, as a representative of thanklessness, since in Chapman we have—

> What's that, a columbine?
> No! that thankless flower grows not in my garden.

Why "thankless" is a question still to be answered. Turner speaks only of the supposed medicinal virtues, Matthiolus and Gerard do no more.

One cannot part from Shakspere's flowers without recalling what he has told us of the little wizards of the

hive. Under Providence, he says, there is special work
in the world for everybody, and the faithful concurrence
of the whole, duly performed, is "like music."

> Therefore doth heaven divide
> The state of man in divers functions,
> Setting endeavour in continual motion,
> To which is fixèd, as an aim or butt,
> Obedience.
>   So work the honey-bees,
> Creatures that, by rule in nature, teach
> The act of order to a peopled kingdom.
> They have a king, and officers of sorts;
> Where some, like magistrates, correct at home;
> Others, like merchants, venture trade abroad;
> Others, like soldiers, armèd in their stings,
> Make boot upon the summer's velvet buds;
> Which pillage they, with merry march, bring home
> To the tent-royal of their emperor,
> Who, busied in his majesty, surveys
> The singing masons building roofs of gold;
> The civil citizens kneading up the honey;
> The poor mechanic porters crowding in
> Their heavy burdens at his narrow gate;
> The sad-eyed justice, with his surly hum,
> Delivering o'er to execùtors pale
> The lazy, yawning drone.—*King Henry the Fifth*, i., 2.

As an account of the domestic economy of the honey-
bees, this most beautiful description, it hardly needs the
saying, is inexact. Bee-life, in the Shaksperean age, had
not been studied. The main idea, however, is faithfully
put, and it is for the reader who knows the particular
truth to correct it mentally as he reads. There are
many other allusions to bees and honey, in all, probably,
not fewer than thirty. *In All's Well*, for instance,

This he wished,
Since I nor wax, nor honey, can bring home,
I quickly were dissolvèd from my hive,
To give some labourers room.—(i., 2.)

Some of them are very beautifully figurative:—

O my love, my wife !
Death that hath suck'd the honey of thy breath,
Upon thy beauty yet hath had no power.
Thou art not conquer'd; beauty's ensign yet
Is crimson in thy lips, and in thy cheeks,
And death's pale flag is not advancèd there.

Romeo and Juliet, v., 3.

Chapter Tenth.

CULTIVATED FRUITS, ESCULENT VEGETABLES, AND MEDICINAL HERBS.

Think'st thou it honourable for a noble man
Still to remember wrongs?—*Coriolanus*, v., 3.
    The rarer action is
In virtue than in vengeance.—*Tempest*, v., 1.

H E interest of the Shaksperean references to the garden fruits and vegetables of his period, and to the herbs grown for medicinal purposes, consists chiefly in the information they afford as to the horticulture of the Elizabethan age, and the value then attached to simple herb-physic. The circumstances amid which these references occur are seldom poetical. They differ altogether from the beautiful surroundings of the flowers. Sometimes they are distinctly unpleasing; and the characters with whom they are identified are often of third-rate position in the drama. For completeness' sake, it is of course proper to recognise these various allusions, bearing in mind, at the same time, that many of them may quite possibly not be Shakspere's own, but the unfair interpolations of actors or transcribers. The associations are usually quite lateral. These fruits and so forth seem to be introduced, not so much because of their merits, as to illustrate the jocularity of the speaker, or to serve as the basis of a pun or a *double entendre.* Shakspere wrote his plays, no doubt, with a view to their being performed before audiences to whom amusement and broad farce were more welcome than poetry and philosophy. He knew that he must supply material for fun, and this he did supremely. He is entitled, nevertheless, to the full benefit of the doubt whether the vulgarities and the impurities may not have been inserted by other hands. No manuscripts of his own, as already said, are in existence. Between the time of the original writing of the dramas and the first printing there was plenty of opportunity for desecration, and the desecration would be carried still further by players who found it to their interest to interpolate expressions adapted to "bring down the house," these getting by degrees into the successive manuscript copies. Pope pointed out, more than one hundred and sixty years ago (his edition of Shakspere having been published in 1721), that the *Romeo and Juliet* of the famous "first folio"—the first collective edition of the plays, issued in 1623 by the two players Heminges and Condell—

contains many mean conceits and ribaldries of which there is no hint in the early single-play editions. It cannot be supposed for a moment that these were inserted by Shakspere as after-thoughts. Surely their introduction must have been after the manner indicated.

That Shakspere had no real and personal love for low quibbles and other such vulgarities, he himself plainly shows us in the *Merchant of Venice*, iii., 5, illustrating, as in many other places, his own private ideas, without an atom of ostentation. Launcelot, old Shylock's coarse and ignorant serving-man, talks rudely to Jessica. While they are conversing, he is interrupted by the entrance of Lorenzo, who, if not one of the foremost of the Shaksperean characters, is distinguished at least for his reverence, his good common-sense, and other qualities of the gentleman. It is to Lorenzo that at a more pleasing time we owe the sublimest burst of reverent poetry in the whole of Shakspere:—

> How sweet the moonlight sleeps upon this bank !
> Here will we sit, and let the sounds of music
> Creep in our ears. Soft stillness and the night,
> Become the touches of sweet harmony.
> Sit, Jessica. Look, how the floor of heaven
> Is thick inlaid with patines* of bright gold !

---

* "Patines, small golden dishes employed in the ritual of the Catholic Church." The stars bear no resemblance to *these*. Surely, as proposed, the better word would be *patterns—i.e.*, designs or beautiful geometrical sketches and methods of arrangement, as set forth in the constellations, Orion, in its season, Cassiopeia, and the sleepless Seven, with all the rest of the host of heaven, "shedding sweet influence."

N

> There's not the smallest orb that thou beholdest
> But in his motion like an angel sings,
> Still quiring to the young-eyed cherubim:
> Such harmony is in immortal souls;
> But whilst this muddy vesture of decay
> Doth grossly close us in, we cannot hear it.

Those wonderful words, also, a little later on—

> The man that hath no music in himself,
> Nor is not moved with concord of sweet sounds,
> Is fit for treasons, stratagems, and spoils.
> The motions of his spirit are dull as night,
> And his affections dark as Erebus;
> Let no such man be trusted.

One cannot be other than predisposed in his favour, and thus, when Launcelot resumes his quibbling talk, the comment, never doubt, is from Shakspere's own heart:—

LAUNCELOT: It is much that the Moor should be more than reason; but if she be less than an honest woman, she is indeed more than I took her for.

LORENZO: How every fool can play upon the word! I think the best grace of wit will shortly turn into silence, and discourse grow commendable in none only but parrots.

Still, like all other quibblers, Launcelot is heedless of the rebuke. He catches at every phrase used by Lorenzo, who, when the silly fellow makes his exit, says to himself,—

> The fool hath planted in his memory,
> An army of good words. And I do know
> A many fools, that stand in better place,
> Garnish'd like him, that for a tricksy word
> Defy the matter.

The same spirit of private and personal contempt for all

that is low, vulgar, sham, and contemptible, is set forth in Hamlet's advice to the players—incomparably more Shaksperean in truthfulness to his own nature than all the offensive utterances put together. Refusing him the benefit of the doubt; conceding that he was himself the author of the un-Shaksperean lines and passages; it may at all events be considered probable that, had he revised with the pen, he would have expunged very much when thinking of posterity, or at least have so modified various expressions, as to enable us to rejoice to-day in Shakspere pure and simple, instead of feeling that the volume in our hands is Shakspere *plus* appendages not of his own heart, but attached in order to tickle a public whose chief wish was to be made laugh.

Whatever may be the true history of these vulgarities and impurities, in looking at passages of any kind in Shakspere that offend, let us never be too severe. The pure and beautiful outweighs the contrary beyond measure. We should so accustom ourselves to rejoice in his wisdom and grace as to forget the uncomely, passing it over just as we ignore the disagreeables of daily life. If we care for a true artist, whatever his department of work, this is the true way, not only to honour, but to enable ourselves to appreciate him, just as in contemplating the works of God, the charm consists in habituating one's-self to companionship with what is loveliest in them and most glorious. People who are sincerely interested in beauty and the pure, do not look for blots and deficiences. They have enough to occupy their time without. The disposition to hunt for flaws in things constitutionally good and noble, before we are quite sure that we have realised all the worth, is a malady to be regarded with special terror. No disorder can be more disastrous to one's intelligence, since the mind must needs become degenerate in the exact ratio of its interest in the defective, which last can never be either nourishing or healthful. The most unenviable of aptitudes is that one which leads people to look for blemishes in things better than themselves.

The garden fruits named by Shakspere are all of the very ordinary kinds, and such as had been in cultivation in this country, in most cases, probably, since the time of the Romans, to whom, in all likelihood, the original introduction of the exotic sorts was owing. That the apple, the pear, the cherry, the plum, and the medlar, exist in England, in the wild condition, is no doubt true; but that the eatable varieties cherished in orchards and gardens in the Shaksperean age had been developed in our own island can hardly be supposed: they would be descendants of imported trees, perhaps of the Norman period, or the Plantagenet. The grape, the fig, the mulberry, the quince, the walnut, the chestnut, the peach, the apricot, it hardly needs the saying, were in the first instance chiefly of oriental origin, moving westwards with wheat and other features of civilisation.

Shakspere turns to the fruit-trees for two or three of his most beautiful pictures of the vicissitudes of human life:—

> Then was I as a tree
> Whose boughs did bend with fruit; but in one night
> A storm, a robbery, call it what you will,
> Shook down my mellow hangings, nay, my leaves,
> And left me bare to wither.—*Cymbeline*, iii., 3.

So in that most profoundly, not to say sacredly pathetic of all his scenes, the story of the downfall of Cardinal Wolsey:—

> So farewell to the little good you bear me.
> Farewell, a long farewell, to all my greatness!
> This is the state of man.   To-day he puts forth
> The tender leaves of hope; to-morrow blossoms,
> And bears his blushing honours thick upon him.
> The third day comes a frost, a killing frost,
> And,—when he thinks, good easy man, full surely,
> His greatness is a-ripening, nips his root,
> And then he falls, as I do.
>   I have ventured,
> Like little wanton boys that swim on bladders,
> This many summers in a sea of glory,
> But far beyond my depth.   My high-blown pride
> At length broke under me, and now has left me,
> Weary, and old with service, to the mercy
> Of a rude stream, that must for ever hide me.
> Vain pomp and glory of this world, I hate ye.
> I feel my heart new-open'd.   O how wretched
> Is that poor man that hangs on princes' favours!
> There is, betwixt the smile we would aspire to,
> That sweet aspèct of princes, and their ruin,
> More pangs and fears than wars and women have;
> And when he falls, he falls like Lucifer,
> Never to hope again. —*Henry the Eighth*, iii., 2.

" Blushing honours " would seem to have been suggested by the incomparable May roseate of the apple-trees, since all other standard fruit-bearers which have petaled flowers, are altogether snowy, and reddening fruit is never liable to be thrown down by frost. Surely, too, it would be the apple-trees, and the pears alongside, which suggested the pleasant comparison of the first sweet shoots of love to opening buds or sprouts, these being what are intended by the Shaksperean "springs" in *Venus and Adonis*, 110,

> This canker that eats up love's tender spring;

in the *Rape of Lucrece*,

> Unruly blasts wait on thy tender spring;

and in the *Comedy of Errors*, iii., 2,

> Even in the spring of love, thy love-springs rot.

The same may be the intent, perhaps, of the phrase in the charming lines, quoted on page 21, from the *Midsummer Night's Dream*, ii., 2:—

> And never, since the *middle summer's spring*,
> Met we on hill, in dale, forest, or mead,
> Or on the beachèd margent of the sea,
> To dance our ringlets to the whistling wind,
> But with thy brawls thou hast disturbed our sport.

Titania is upbraiding Oberon and complaining of his revels. That they have been married for many years may safely be assumed. The present quarrel, a "very pretty" one, is certainly not the first in which the royal couple have been engaged. From this point of view the "middle summer's spring" may perhaps refer to a period when love between them was more abounding, and there were fewer of the "forgeries of jealousy." There are difficulties, however, in the way of thus understanding the phrase such as do not interfere with the interpretation given above. "Orchard," we must remember also, in the Shaksperean times, was often equivalent to "garden," as in *Much Ado*, ii., 3,

> The orchard-walls are high and hard to climb;

and in *Romeo and Juliet*,

> He ran this way, and leaped this orchard wall.—(ii., 1.)

The APPLE itself, of which literature of every kind has always been so fond, holds quite a subordinate place in regard to prettiness of the allusions—about a dozen altogether, — most of them being allied to the jocular, and scarcely one distinctly elegant. Usually it is a named variety that is cited, as the "pomewater," in *Love's Labour's Lost*, iv., 2; the "leathercoat," in *2nd Henry the Fourth*, v., 3; the "pippin," in the *Merry Wives*, i., 2 ; the "bitter-sweeting," in *Romeo and Juliet*, ii., 4; the "costard," a large and coarse variety that served as a contemptuous metaphor for a vulgar and ignorant head, as in *Lear*, iv., 6, and elsewhere; and the John-apple, or "apple-john," a fruit of good flavour, and that kept good for two years, but naturally during that time became very shrunken, so that old Falstaff's comparison was well based, and his dislike of it quite intelligible (*1st Henry the Fourth*, iii., 3; *2nd Henry the Fourth*, ii., 4). More pleasing allusions occur in *Twelfth Night*, v., i.:—

> An apple cleft in two is not more twin
> Than these two creatures;

and in the *Tempest*, ii., 1, "I think he will carry the island home, and give it his son for an apple." When, in Sonnet xciii., we have—

> How like love's apple doth thy beauty grow,
> If thy sweet virtue answer not thy show,

the reference is obviously to the conventional use of the apple in Scriptural interpretations, to denote the "forbidden fruit" of the Garden of Eden.

The PEAR furnishes Falstaff with a simile not unlike the previous one *(Merry Wives*, iv., 5), and which is almost echoed by Parolles *(All's Well*, i., 1*)*. Mercutio's graceless allusion to the "Popering pear" is rendered interesting by the association of this variety with the village near Calais, of which Leland, the famous antiquary, was some time rector, and from which it was named. When in the *Winter's Tale*, iv., 2, we read of "warden pies," the reference is again to a particular variety, golden yellow when quite ripe, famous for its long-retained goodness, whence the name—A.S. *wearden*, to preserve—and greatly esteemed, not for what we call "pies" to-day, but for the analogues of apple-dumplings.

The QUINCE is mentioned in *Romeo and Juliet*, iv., 4, when Lady Capulet is giving orders for the wedding-feast,

> They call for dates and quinces in the pastry,

the "pastry" being the place in which "paste" was made, as in Nicholas Brereton, 1582,

> Now having seene all this, then you shall see, hard by,
> The pastrie, meale-house, and the roome whereas the coales
> do ly.

The fame of the quince reaches back to the remotest antiquity. There can be no doubt that it was included in the *tappūach* of the Hebrews (in the A.V. rendered "apple"), and that it is immediately intended by the Greek μῆλον and the Latin *malum*. In Theocritus this downy and fragrant fruit supplies a very celebrated metaphor, which no doubt well pleased Boccaccio, and is imitated by Ben Jonson, who calls it the melicoton. It appears in the famous myth of the golden apples of the Hesperides. For the same reason, when Hero became the wife of Zeus, the divinities presented her with appropriate bridal gifts,—Gaia,—the earth personified, providing a tree which bore "golden apples." There is no occasion here to go into the significance minutely. Shakspere is thought to have been no stranger to it, and to have had it in view in the *Romeo and Juliet* passage, which is possible, but not likely, since the esteem in which the fruit was held in his age, for the making of marmalade, or "cotiniac," is quite enough to account for the introduction of it. Miss Wood relates an amusing anecdote as to the regard in which it was held by fickle and insatiable Henry the Eighth. In 1539, the new queen, Anne of Cleves, desired to engage a maid-of-honour. Lady Lisle, seeking to propitiate his majesty in favour of her daughter Katharine, sent him a present of damson-cheese, and some of this identical cotiniac. Whether the object was attained or not, we are left in doubt. So acceptable, however, to the royal epicure were Lady Lisle's sweetmeats, that Anne Bassett, by whose hand they had been conveyed, writes off-hand, "The king doth so like the conserves you sent him, that his grace commandeth me to send to you for more, and this as soon as may be."

The MEDLAR. Partly, no doubt, because of its un-inviting appearance when sufficiently softened to be passed across the lips; partly because of the very peculiar astringency, the medlar has never held a high place among the orchard fruits; and almost invariably, when mentioned in literature, the associations are ignoble, if not offensive. Apemantus cites it in *Timon of Athens*, playing jocularly upon the sound of the name,—

> There's a medlar for thee, eat it.
> TIMON: On what I bite, I feed not.
> APEMANTUS: Dost hate a medlar?
> TIMON: Ay, though it look like thee.
> APEMANTUS: And thou had'st hated meddlers sooner, thou
>   should'st have loved thyself better now.
>
>   (iv., 3.)

Lucio also mentions it, in *Measure for Measure*, iv., 3, and Mercutio, whose gay laugh "rings down the street," in *Romeo and Juliet*, ii., 1. Rosalind comes just in time to atone for the latter, bantering old Touchstone, and again with play upon the double sense of the word, though we must be careful not to mistake her intent in calling the medlar the "earliest fruit of the season." She does not mean that it is the first to be gathered from the bough, but the first of the hard ones collected for winter store which after gathering becomes soft enough to eat.

The PLUM. The allusions to the plum, like those to the medlar, both in Shakspere and in literature generally, are, one with another, upon a very low level; at all events, they make little appeal to one's intellectual tastes. The tree is mentioned in *Venus and Adonis,* in the *Passionate Pilgrim,* and in the farcical scene in *2nd Henry the Sixth,* ii., 1, where also the fruit is named. The fruit appears again in the *Merry Wives,* v., 5, " I will dance and eat plums at your wedding." Many varieties have been in cultivation from very early times. The *Henry the Sixth* passage includes notice of the damson, or damascene, attributed to the gardens of ancient Damascus. Prunes, the dried produce of a sort grown chiefly in France, are mentioned upon no fewer than five occasions, but never once invitingly. Happily, there is once again a set-off, though the scene in which it occurs is one of those which make the heart beat with mingled sympathy and indignation. Poor Constance, the deceived, the friendless, lies at the mercy of bitter-tongued Elinor, little Arthur, "my fair son," guessing, only too sadly, what it all means:—

ELINOR: Who is it thou dost call usurper, France?
CONSTANCE: Let me make answer—thy usurping son.
ELINOR: Out, insolent! Thy bastard shall be king;
            That thou may'st be a queen, and check the world!
CONSTANCE: My bed was ever to thy son as true,
            As thine was to thy husband; and this boy
            Liker in feature to his father Geoffrey,
            Than thou and John in manners; being as like
            As rain to water, or devil to his dam.
            My boy a bastard! By my soul I think,

His father never was so true begot;
It cannot be, an if thou wert his mother.

ELINOR: There's a good mother, boy, that blots thy father?
CONSTANCE: There's a good grandam, boy, that would blot *thee!*

.   .   .   .   .

ELINOR: Come to thy grandam, child !
CONSTANCE: Do, child.   Go to it grandam, child :
            Give grandam kingdom, and it grandam will
            Give it a plum, a cherry, and a fig :
            There's a good grandam ! *—*King John,* ii., 1.

One can see the quivering lips, the red lids, the hot tears coursing fast down the pallid cheeks.   By-and-by comes the infinite, yet still most queenly grief.   Arthur, as we all know, perishes cruelly: "An image," says Mrs. Jameson, "more majestic, more wonderfully sublime, was never presented to the fancy :—

I will instruct my sorrows to be proud;
For grief is proud, and makes his owner stout.
To me, and to the state of my great grief
Let kings assemble; for my grief's so great
That no supporter but the huge firm earth
Can hold it up.   Here I and sorrow sit;
Here is my throne—bid kings come bow to it.
            Throws herself on the ground.—(iii., 1.)

"Not only," continues that accomplished authoress, "do her thoughts start into images; her feelings become persons; grief haunts her as a living presence :—

---

* Observe, in the above passage, the very interesting old-fashioned employment of *it* where now we say *its,* which last, when Shakspere wrote, was but slowly coming into use.   With many other Shaksperean archaisms, this form of the word is still extant in the Lancashire dialect;—no phrases are more common than such as "Come to *it* mammy."

> Grief fills the room up of my absent child,
> Lies in his bed, walks up and down with me;
> Puts on his pretty looks, repeats his words,
> Remembers me of all his gracious parts,
> Stuffs out his vacant garments with his form;
> Then have I reason to be fond of grief.—(iii., 4.)

The CHERRY. The Shaksperean references to the cherry, introduced by the above in *King John*, contrast strongly with those to all the other fruits yet mentioned, being in every instance pure, if not breathing the genuine spirit of poetry. How pretty the picture in *Venus and Adonis*, one of the loveliest ever drawn of reciprocity of goodly service:—

> When he was by, the birds with pleasure look,
> That some would sing, some other in their bills
> Would bring him mulberries and ripe red cherries,
> *He* fed them with his sight; *they* him with berries.

Then comes the fond comparison in the *Midsummer Night's Dream*, when Demetrius talks of Helen:—

> O, how ripe in show,
> Thy lips, those kissing cherries, tempting grow!—(iii., 2.)

No fruits are individually more like one another than cherries, or more apt to conjoin. Hence in *Henry the Eighth*, v. 1:—

> 'Tis as like you,
> As cherry is to cherry;

and in the *Midsummer Night's Dream*, iii., 2, that exquisite delineation by Helena of the mutual love between herself and Hermia in the days of their un-chequered girlhood, now so sadly marred. If one

sentiment more than another was specially dear to Shakspere, it was the sanctity of friendship:—

> O, and is all forgot?
> All schooldays' friendship, childhood innocence?
> We, Hermia, like two artificial* gods,
> Have with our neelds created both one flower,
> Both on one sampler, sitting on one cushion,
> Both warbling of one song, both in one key;
> As if our hands, our sides, voices, and minds,
> Had been incorporate. So we grew together,
> Like to a double cherry, seeming parted,
> But yet a union in partition;
> Two lovely berries moulded on one stem;
> So with two seeming bodies, but one heart;
> Two of the first, like coats in heraldry,
> Due but to one, and crownèd with one crest.
> And will you rend our ancient love asunder,
> To join with men in scorning your poor friend !
> It is not friendly, 'tis not maidenly;
> Our sex, as well as I, may chide you for it,
> Though I alone do feel the injury.

The cherry appears also in the description of the accom-
plishments of Marina, in *Pericles:*—

> She sings like one immortal, and she dances
> As goddess-like to her admirèd lays;
> Deep clerks she dumbs, and with her neeld composes
> Nature's own shape, of bud, bird, branch, or berry,
> That even her art sisters the natural roses,
> Her inkle, silk, twin with the rubied cherry.
>
> (Act v., proem.)

The "deep clerks" who failed in discussion with the
royal Tyrian maid, were, in the Shaksperean sense of the

---

* "Artificial," here curiously used in the sense of skilful or
ingenious.

words, learned men, or such as to-day would be termed "scholars," the fine old original meaning of "clerk" being in the Elizabethan age still the current one. A more interesting word is not to be found in the English language, since it takes us back to the time of the Druids, and is coeval with the most ancient Celtic monuments. The priesthood in those times was three-fold, the second of the three orders, called the Bards, having for their special office the composition of verses, which they sang to the music of the harp. The harp was called *clar;* the bard was a *clarsair;* that which related to the harp was *clarach.* Taken up by the Romans, this word was Latinized into *clericus.* The Bards or harpers were men of high attainments for their period;—*clericus* thus came to signify mental culture, and a "clerk" one who was better informed than the people around him.

A favourite game with the boys of Shakspere's time was "cherry-pit," often alluded to in the contemporary literature, and by himself in *Twelfth Night,* iii. 4.

The Fig. Excepting in the *King John* passage, ii. 1, and the verses in the *Midsummer Night's Dream,* where Titania gives the injunctions that in performance would be so gratifying,—

> Feed him with apricocks and mulberries,
> With purple grapes, green figs, and mulberries.

Excepting in these two places, the Shaksperean allusions to the fig have no attractions. From time immemorial, this fruit has been a metaphor for the worthless and the

mean. Why so, there is no occasion to inquire too
curiously. Light enough is cast upon it by the classical
poets. Shakspere quite appropriately introduces the fig
in similar association, not forgetting the employment of
it to express insult and contempt, and more than this
need not be said.

The APRICOCKS of the lines last quoted are the
"apricots" of to-day, by Turner called the "hasty peche."
Early as they are to arrive, apricots at midsummer would
in England still be a phenomenon. This does not matter
just now. The scene is laid in fairy-land, where times and
seasons are indifferent, and in the climate of the Graces.
Another allusion to this fruit occurs in *Richard the Second*,
iii., 4, already cited (p. 150). The peach *ipsissima*,
though well-known in Elizabethan gardens, is adverted to
by Shakspere in reference simply to its colour.

GRAPES are mentioned in the Shaksperean dramas
upon ten occasions. The beautiful climbing shrub, with
its green tendrils, which produces them—the vine—is
noticed in just as many other places; the vineyard
receives notice nearly as often. There is no instance
of anything distasteful in the grape allusions; the poetic
passages are nevertheless those which refer rather to
the plant, foremost among them coming Cranmer's fine
paraphrase of Scripture during his benedictions at the
baptism of Elizabeth:—

> In her days every man shall eat in safety
> Under his own vine—what he plants, and sing
> The merry songs of peace to all his neighbours.
>
> *Henry the Eighth*, v., 5.

While the delicious masque in the *Tempest,* iv., 1, is in progress, Iris speaks of the " pole-clipt vineyard." Here we are treated to the employment of one of the capital old Anglo-Saxon words now forgotten except among the rustics in particular districts — *clyppan,* to embrace. Trained, as was formerly the practice with grape-growers, so as to encircle poles, the vine literally clasps or embraces them. " Our king, being ready to leap out of himself for joy of his found daughter . . . again worries he his daughter with clipping her" (*Winter's Tale,* v., 2). In Lancashire, to this day, to twine the arms fondly around one is to "clip."

Several fruits come in for trifling or incidental allusion, as the GOOSEBERRY, in *Second Henry the Fourth,* i., 2; the STRAWBERRY, in *Othello,* iii., 3; *Richard the Third,* iii., 4; and *Henry the Fifth,* v., 1.; the WALNUT, in the *Merry Wives,* iv., 2, and the *Taming of the Shrew,* iv., 3; and the CHESTNUT in the same play, i., 2; *Macbeth,* i., 3; and *As You Like It,* iii., 4, where Celia adduces the colour of the ripe shell as corresponding with that of Orlando's glossy hair. The MULBERRY, the last in the list, had long been established in English gardens when Shakspere wrote. His personal regard for this famous tree led, it hardly needs the saying, to the planting by his own hand, of one which, but for sacrilege, would probably have been a joy to this very day, since the mulberry is one of the longævals. The fruit is referred to in *Venus and Adonis, Coriolanus,* iii., 2; and in the *Midsummer Night's Dream,* iii., 1. In v., 1 of the same play we have the beautiful

o

picture of Thisbe "tarrying in the mulberry shade," so soon to be exchanged for the turf and the throbbing moment when, the moon high in the heavens, and the "sweet wind did gently kiss the trees," she herself, panting, did

> Fearfully o'ertrip the dew,
> And saw the lion's shadow ere himself,
> And ran, dismayed, away.

Alas, little veil!—how much thou hast to answer for in the story of human love!

THE ESCULENT VEGETABLES.

The esculent vegetables mentioned by Shakspere—those cultivated in the kitchen-gardens of the time,—are adduced like most of the fruits, chiefly for some kind of facetious or jocular use, and only a few of them appear more than once or twice. The history of the introduction to this country of the sorts indigenous to other lands, the times when brought, and by whom, constitutes a very interesting chapter in the annals of horticulture. Here it is sufficient to note that several of the best of our cultivated vegetables, like many of the best cultivated fruits, are accounted native productions of old England, though developed, probably, into the condition in which we now have them upon the continent. Such are the turnip, the carrot, which seems to be owing to the skill of the Flemings of the time of Edward III., and the immensely diversified plant still represented in the wild cabbage of our sandy shores. The very ancient kinds, such as peas, broad or Windsor beans, the radish, and the lettuce, were in all likelihood, introduced by the Romans. The first-mentioned, the turnip, appears in the *Merry Wives*, iii., 2, which play also contains the sole allusion to the carrot (iv., 1). Peas, Biron tells us, are good not only for man, but for birds:—

> This fellow picks up wit as pigeons' pease.
> *Love's Labour's Lost*, v., 2.

The young green shells are also referred to under their obsolete name of "squashes," derived from the onomato-pœtic French *esquacher*, or Italian *squasciare*, as in the *Midsummer Night's Dream*, iii., 1, and in *Twelfth Night*, i., 5, "Not yet old enough for a man, nor young enough for a boy, as a squash is before 'tis a peascod." "Peascod" was the appellation of the matured pod, or when ready for gathering; and after the manner of many an ancient and beautiful intimation in colloquial speech of the season intended, without actually mentioning it, by citation of some characteristic product, — in *2nd Henry the Fourth* is made a synonym of summer,—"Well, fare thee well! I have known thee these twenty-five years come peascod-time." Touchstone, in *As You Like It*, ii., 4, refers to a somewhat celebrated old method of divination practised by lovers with the pods. A fully ripe one was selected and snatched away suddenly from the stem. If it broke, and the peas were scattered, the omen was bad: if it remained whole, the promise was happy,—though in the present instance, of Jane Smile and "her pretty chopped hands," we hear no more. Beans are mentioned in *1st Henry the Fourth*, ii., 1, in the same line with another reference to peas.

That various forms of the *Brassica oleracea* have been cultivated from time immemorial, seems indubitable. In Shakspere, however, the reference is collective:—"Good worts," exclaims old Falstaff, catching at Evans' faulty pronunciation, "worts" instead of "words," and playing upon it,—"Good worts ! Good cabbages !" *(Merry Wives*, i., 1.) With Sir John this term was plainly equivalent to the modern market-woman's "greens." Etymologically, the word is primæval. It appears in Ælfric's translation of Genesis ii., 5., "and eall gærs and wyrta ealles eardes." Then in Chaucer, when the fox lies concealed, biding his time to pounce upon chanticleer. In Baret, 1580, we have "wourts, all kind of herbes that serve for the potte," cole or kail wort included. Aristophanes, says Lyte, "taunted his Euripides with the remark that his mother was not a seller of wurtes or good pot-herbs, but only of scandix." It is the same, with still further extended significance, which appears in the plant-names star-wort, stitch-wort, rib-wort, glass-wort, cross-wort, sneeze-wort, and fifty others.

In what degree of estimation the Leek was held when Shakspere wrote, does not appear with any clearness, and the poet's own allusions to it are only lateral. The beautiful emerald shade of the leaves adverted to in *chrysoprasus*, as well known, is very anciently proverbial; hence it is aptly cited in the comic parody of the

Pyramus and Thisbe story in the *Midsummer Night's Dream*, v., 2 :—

> His eyes were green as leeks ;—
> O Sisters three
> Come, come to me!

The "Sisters" invoked are the classical Fates;—

> Since you have shore
> With shears his thread of silk,
> Tongue, not a word ;—
> Come, trusty sword ;
> Come, blade, my breast imbrue ;
> And farewell, friends,
> Thus Thisbe ends,
> Adieu, adieu, adieu!

The other interesting association—that of the leek with primitive Welsh history, is also placed before us, appearing several times in *King Henry the Fifth* (iv., 1 ; iv, 7; and v., 1, in the encounter between Pistol and Fluellin). The origin of the use of this plant as the Welshman's emblem is referred to a period as early as the days of King Arthur, whose uncle, St. David, having gained a great victory over the Saxons, commanded his troops thenceforward to wear, every one of them, a leek upon the anniversary of his death, which occurred March 1st, A.D. 550. It is to be feared, however, that the specific identity of the plant intended by his ancient majesty is quite as problematical as that of St. Patrick's shamrock,*

---

* The national badge of Ireland is generally thought to be the white clover, *Trifolium repens.* But the nonsuch, *Medicago Lupulina*, is also honoured, and so, in truth, are several other little trefoils. None of these, after all, meet the requirements, which are of a leaf

since in its earliest signification, leac denoted any kind of green and juicy plant, as still indicated in the compounds houseleek, hemlock, and charlock. It may be worthy of notice here, that the Hebrew word *chātsir,* translated "leeks" in Numbers, xi., 5, is elsewhere in the Old Testament rendered "grass." "He causeth the *chātsir* to grow for the cattle" (Psalms, civ., 14). "Behold now behemoth which I made with thee; he eateth *chātsir* as an ox (Job, xL, 15).

The Onion, like the leek, is never actually spoken of by Shakspere as an esculent, though mentioned as tainting the breath *(Midsummer Night's Dream,* iv., 2), and several times in reference to the effect of the vapour upon the eyes, drawing tears. In the *Taming of the Shrew* this is used satirically for sham or pretended grief:

> And if the boy have not a woman's gift
> To rain a shower of commanded tears,
> An onion will do well for such a shift.—*(Introd.)*

Again in *Antony and Cleopatra,* i., 2, "Indeed the tears live in onion that should water this sorrow;" while in *All's Well,* v., 3, it becomes a figure for mental perception of coming trouble,—not an elegant one, it must be acknowledged, though appropriate to the occasion,—

> Mine eyes smell onions, I shall weep anon.

Garlic, very naturally, appears only in similar association, or as an article of food such as the vulgar alone would care for—*Midsummer Night's Dream,* iv., 2; *Measure*

---

constituted of three exactly similar leaflets, a form supplied, in the British islands, solely by the wood-sorrel, *Oxalis Acetosella.*

for Measure, iii., 2; *Coriolanus*, iv., 6; *1st Henry the Fourth*, iii., 1. The Radish is mentioned in the same play, ii., 4, and again in the Second Part, iii., 2. The Lettuce appears in *Othello*, i., 3, where also there is a reference to garden Thyme. Parsley is mentioned in a passage one does not care to re-peruse (*Taming of the Shrew*, iv., 4). Salad-plants in general are alluded to under the name of "grass" by Jack Cade (*2nd Henry the Sixth*, iv., 10, twice); and by the clown in *All's Well*, iv., 5—"I am no great Nebuchadnezzar, sir; I have not much skill in grass." The famous Chaldæan monarch's "grass," it is hardly necessary to say, was not grass in the botanical sense of the word, but green herbaceous vegetation, in the Hebrew, *'āṣābh*.

Excepting the pumpion or pumpkin, which would seem to have been grown in the Elizabethan gardens as a sort of ornamental adjunct, and which is used in the *Merry Wives*, iii., 3, as in old Greek literature, metaphorically, for a great dullard or empty-pate;—excepting this, the Shaksperean vegetables comprise, in addition, only a few of the class called seasoning-herbs. Perdita mingles mint, savory, and marjoram with her marigolds (*Winter's Tale*, iv., 3). Marjoram appears also, not very intelligibly, in Sonnet xcix; and in *All's Well*, iv., 5:—"Indeed, sir, she was the sweet marjoram of the salad." The *Lear* passage, iv., 6, introduces the name simply as a watchword. Fennel possesses more interest, being associated both with Falstaff and with Ophelia. Poins, says the dissolute and sensual, though jovial old knight, "plays at quoits well, and eats conger and fennel" (*2nd Henry the Fourth*, ii., 4), a habit quite as probably his own, and fashioned not more upon the flavour of the plant, when used as a condiment, than upon a superstition of the time which Falstaff, at least, would not care to quarrel with. "Auttours," says Turner (1568), "wryte that serpentes waxe yonge agayne by tastinge and eatynge of this herbe, wherefore sum thinke that the use of the herbe therefore is very mete for aged folke" (vol. 2, p. 5). When Laertes' unhappy sister, addressing the king, says, "There's fennel for you, and columbines," there seems to be an allusion to another ancient belief, mentioned by Pliny, and adopted by many writers of the Shaksperean age, namely, that fennel improves the eyesight. "It hath a wonderful propertie to take away the film or web that overcasteth and dimmeth our eyes." She hopes, by the gift of it, to quicken the royal consciousness, just, as a minute before, thinking of Hamlet and her lost bridals, her lips play with rosemary, the emblem of remembrance. Fennel, in the Elizabethan age, was also used as an emblem of flattery:—"Little things catch light minds, and fancie (love) is a worm that feedeth first upon fennel." * One cannot but remember, when in the presence of these green succulents, the happy metaphor in *Antony and Cleopatra*, i., 5:—

> My salad days,
> When I was green in judgment.

---

* Lyly. *Sappho*, ii., 4.

. ODORIFEROUS AND MEDICINAL HERBS.

The home gardens, when Shakspere wrote, appear to have been rich in aromatic or odoriferous herbs and under-shrubs, many of which held a conspicuous place in medicine. They were the store-houses whence the good wives and village dames drew their supplies for the manufacture, chiefly by distillation, of cordials and simple domestic physic. In an age when the use of mineral drugs had scarcely commenced, and faith in "potions" was unbounded, the little plots of balm, rue, and horehound were almost sacred. The tradition was still alive of virtues in plants equivalent to the miraculous,

> In such a night,
> Medea gathered the enchanted herbs,
> Which did renew old Æson;

the "Botany" of the times itself consisted very largely of search into, and enumeration of their qualities; fable was even more rife than the reality. Hence, Ophelia's employment in the passage just referred to, of Rosemary, that beautiful azure-blossomed shrub which the ancients associated with the spray of the sea, whence the name, *ros marinus*, literally "sea-dew," as in that fair old picture where care of the toilet and conjugal affection run, as they always should do, hand in hand;—"Her hair is smoothed with a comb; now she decks herself with rosemary; again with violets or roses; sometimes wears white lilies; washes twice a day her face in springs that trickle from the top of the Pagasæan wood; and

twice she dips her body in the stream." *   The particular
virtue ascribed to it was that of strengthening the
memory.   "Rosemarie comforteth the brayne, and re-
storeth speech; especially the conserve made of the
flowers thereof with sugar." †   Spenser calls it "refreshing rosmarine."   Newton, in the Bible Herbal, says it
"recreateth and cheereth both the heart and mind
of man."   Turner adds that "men do put rosa-mary in
medicines that dryve werisumnes away."   In the suave
little poem, "a Nosegay, alwaies sweet, for Lovers," in
the "Handefull of Pleasant Delites," already quoted
(p. 81), we have—

> Rosemarie is for remembrance
> Betweene us daie and night,
> Wishing that I might alwaies have
> You present in my sight.

Can we wonder that it was consecrated to friendship, as
illustrated so tenderly in Sir Thomas More; and that
in the realm of poetry it should be the first to be offered
by courteous Perdita:—

> Give me those flowers, there, Dorcas.   Reverend sirs,
> For you there's rosemary, and rue; these keep
> Seeming and savour all the winter long.
> Grace and remembrance be to you both,
> And welcome to our shearing !—(*Winter's Tale*, iv., 3.)

It was introduced, for the same reason, among the
symbols used at funerals, as in *Romeo and Juliet*, iv., 5,

> Dry up your tears, and stick your rosemary
> On this fair corse;

and in the spirit of the gold ring, emblem of constancy,

---

* Ovid, Met., xii., 409-413.   † Lyte, p., 264.

among those used also at weddings, when a sprig of rosemary was put in the wine dedicated specially to good wishes for the bride's happiness. Rosemary is mentioned also in *King Lear*, ii., 3; in *Pericles*, iv., 6; and, facetiously, in *Romeo and Juliet*, ii., 4.

RUE, distinguished for its repulsive odour, was also celebrated in the bygones as a plant able to invigorate the memory. "The juice, with vinegar," says Lyte, "given to smell unto, doth revive and quicken such as have the forgetful sicknesse." Shakspere mentions it upon five occasions, but never in the strictly literal sense. Without any etymological connection, it so happened that while the Latin name of the plant was *ruta*, regret and remorse were in old English *ruth*, while the verb denoting these emotions was, as it continues to the present day, "to rue."

Playing upon the similarity of terms, rue was made by our forefathers, always glad of such an accident, the emblem of sorrow and repentance. The transition was easy to the idea of Grace, in the Scriptural sense of the word; and at last the plant itself was called "Herb o' Grace," and simply "Grace." The Divine favour is the most excellent of benisons; for this reason, it is invoked by Perdita. Ophelia, on the other hand, regards the plant as the emblem simply of grief. Hence while giving a portion of what she holds to the queen; the remainder she takes to herself—"There's rue for *you*, and there's some for *me*" (in tone, alas, how piteous!) "We may call it herb o' grace on Sundays;—you may

wear *your* rue with a difference." Why "on Sundays"
does not appear. Perhaps it was a current phrase or
saying of the time.* "With a difference" is more intelligible, the words being borrowed from the language
of heraldry, though by no means correctly used in the
present instance, the heralds' "differences" (the crescent,
the martlet, the annulet, &c.) denoting seniority of sons
when there are several, whereas Ophelia means that with
the queen it should be the emblem of something more
than simple grief,—contrition in regard to the past. The
heraldic phrase occurs also in *Much Ado*, i., 1, when
"my dear lady Disdain," satirical, aggravating Beatrice,
who, with all her airs and graces, is still a true, tender-
hearted, loving woman, says of Benedick, "So now, if
he have wit enough to keep himself warm, let him bear
it for a difference between himself and his horse." She
means nothing unkind,—since for one who is a Beatrice
in capacity for sincere affection to be unkind is quite
impossible. The other Shaksperean allusions to rue
occur in *Richard the Second*, iii., 4; *All's Well*, iv., 5;
and *Antony and Cleopatra*, iv., 2,—

> Grace grow where these drops fall !

---

* For various conjectures as to the meaning, see *Notes and Queries*,
March 10th, 1883, p. 193. Among them is a quotation from
Dr. Warburton:—"Rue was a principal ingredient in the potion
which the Romish priests used to force the 'possessed' to swallow
when they exorcised them. These exorcisms were generally
performed upon a Sunday, in the church, before the whole congre-
gation." Contracted into "Herbygrass," it was stated in *Notes and
Queries*, Nov. 18th, 1882, that the name is still in use near Sheffield.

WORMWOOD, adverted to by the old nurse in *Romeo and Juliet*, i., 3, and in *Hamlet*, iii., 2, in both instances with reference to its bitterness, serves also in *Love's Labour's Lost*, v., 2, for an expressive metaphor,—

> To weed this wormwood from your fruitful brain,—

this last use recalling the frequent introduction of the same figure in Scripture, "the wormwood and the gall," affliction the most severe. The literature of all later ages supplies similar examples. In the melancholy Pontic Epistles of the unfortunate author of the *Metamorphoses* (in which Epistles, by the way, there is scarcely a reference to trees and flowers), the "unsightly plains" produce no more than the bitter unhappy wormwood" (3, i., 23, and again in 3, viii., 15). Horace seems to put wormwood for *physic*, the reputation of which has from the beginning been that of bitterness.—*Epistles*, 2, i., 14.

HYSSOP appears in *Othello*, i., 3; LAVENDER in the *Winter's Tale*, iv., 3; BALM (the *Melissa officinalis*), in *Antony and Cleopatra*, v., 2; and again in the *Merry Wives*, v., 2,—

> The several chairs of order, look you, scour
> With juice of balm, and every precious flower.

The last passage recalls pleasantly the description of good old Baucis when, preparing for her guests, she cleanses and sweetens her cottage table with "green mint," with those other beautiful usages of antiquity when resort was made for similar purposes to verbena. In these charming echoes of the past, most probably quite unperceived by the poet himself, consist not a few of the fascinations that give him immortality. Like the tips of sea-shells half embedded in the brown ripples of the still wet sand, such allusions tell of far more awaiting research. They are gentle hints and invitations, with sure reward beyond. Finally, there is the CHAMOMILE, introduced in *1st Henry the Fourth*, ii., 4, where Shakpere's object seems to be, for once, to make merry at the expense of a writer of his own period, or nearly so,— John Lyly, author of *Euphues, or the Anatomie of Wit*, originally published in 1523. Crowded with pedantry and affectations, injurious, therefore, in its influence upon the literary style of his admirers, this celebrated work was, in its way, nevertheless, moral, and of good practical service, becoming with the courtiers of the time a sort of guide to stately manners. Upon it rose the school of the Euphuists, writers who delighted in puns, alliteration, and antithesis. "Though camomile," says Lyly, the more it is trodden and pressed down, the more it spreadeth" (a well-known fact in botany); "yet the violet, the oftener it is handled and touched, the sooner it withereth and decaieth." "Harry," says Falstaff, "I do not only marvel how thou spendest thy time, but also how thou art accompanied; for though the camomile the more it is trodden on, the faster it grows, yet youth, the more it is wasted, the sooner it wears." Whatever may have been intended with regard to Lyly, remember that these latter words of wisdom are from the same heart that spoke in Adam:—

All this I give you.   Let me be your servant.
Though I look old, yet am I strong and lusty,
For in my youth I never did apply
Hot and rebellious liquors in my blood;
Nor did not with unbashful forehead woo
The means of weakness and debility.
Therefore my age is as a lusty winter,
Frosty, but kindly.—*As You Like It*, ii., 3.

Several of the above-named garden herbs are accounted indigenous to the south of England, where, especially upon sea-side cliffs, and in sandy pastures where there is smell of the waves, parsley, fennel, chamomile, and wormwood are at all events thoroughly naturalised. The others, rue, hyssop, lavender, balm, and marjoram are, like rosemary, natives of the south of Europe. When introduced to this country is uncertain. Some think not very long before Shakspere's own time.

Fennel, *Fœniculum vulgare*, is one of the species which differ from the mass of its great family, the Umbelliferæ, in possessing bright yellow flowers, by which, taken in connection with the stature, and the deep green leaves, split into innumerable capillary segments, it may always be easily recognised. Wormwood, *Artemisia Absinthium*, is a curious little bushy plant, covered in every part with greyish and silky down, the flowers forming innumerable half-pendulous and yellowish buttons the size of peas, and borne in a rather close panicle. Rue, *Ruta graveolens*, is also grey in every part, but not downy, the four-petaled yellow flowers very curiously and prettily crimped. Hyssop is told from its near allies among the

Labiatæ, by its spikes of rich violet-purple flowers, varying sometimes to pink. Balm, as regards beauty of aspect, is the least effective of any of the series, the flowers being white and insignificant, and half-concealed among the dark-hued foliage. It must not be confounded with the source of Balm-of-Gilead, mention of which is also made by Shakspere, and upon various occasions, as will be noticed in due course. How the name came to be extended to a plant relatively of such little importance as the melissa, does not appear. It is one of the bequests of the fanciful and conjectural botany of the middle ages.

Chapter Eleventh.

THE FARM.

Silence is the perfectest herald of joy:—I were but
little happy if I could say how much.—*Much Ado*, ii., 1.

HE Shaksperean references to farm-plants,
after the same manner as those to garden
esculents, are interesting chiefly in the
light they reflect upon the agriculture of
the period.   Some are of the kind that
belong to every age and country, the greater portion for
instance of the allusions to grass:—

When Phœbe doth behold
Her silvery visage in the watery glass,
Decking with liquid pearl the bladed grass.
Midsummer Night's Dream, i., 1.

Mowing like grass,
Your fresh-fair virgins and your flowering infants.
Henry the Fifth, iii., 3.

P

These tidings nip me, and I hang the head
As flowers with frost, or grass beat down with storm.
                              Titus Andronicus, iv., 4.

Say to her, we have measured many miles
To tread a measure with her on this grass.
                              Love's Labour's Lost, v., 2.

Now and then, in allusion to the same, some pretty old superstition or rural custom comes in for notice, as at the opening of the *Merchant of Venice,* when Antonio is telling his friends of his anxieties:—

ANTONIO: In sooth I know not why I am so sad;
          It wearies *me;* you say it wearies *you.*
SALARIO: Your mind is tossing on the ocean;
          There, where your argosies with portly sail,
          Like signiors and rich burghers of the flood,
          Or, as it were, the pageants of the sea,
          Do overpeer the petty traffickers,
          That curt'sey to them reverence
          As they fly by them with their woven wings.
SALANIO: Believe me, sir, had I such venture forth,
          The better part of my affections would
          Be with my hopes abroad.   I should be still
          Plucking the grass to know where sits the wind.

A more primitive mode of forecasting the weather it would be difficult to find, unless, perhaps, in the contemplation in the olden time of the scarlet pimpernel, the "shepherds' weather-glass." The allusion is very appositely introduced, the idea being that in times of profound anxiety, men turn to the most trivial signs and tokens of the possible future.   Roger Ascham did not discredit it.   "This way," says he, in his famous old treatise upon Archery, 1571, "I used in shooting.

Betwixt the markes was an open place. There I take a fethere, or a lyttle grasse, and so learned how the wind stood."

Shakspere's references to grass are probably not fewer than twenty or thirty, but he never means anything precise as to kind. No term applied to the vegetable productions of the soil is used in a broader sense. The number of genuine botanical grasses in England considerably exceeds a hundred, and of these about thirty help to supply the natural food of sheep and cattle, though not more than fifteen different kinds ever occur in the same meadow. Mingled with them there are scores of little plants which, botanically, are not grasses at all. The word is to be taken, accordingly, in Shakspere as it is colloquially, or as denoting the components, whatever they may be, of the turf, the meadow, and pasture of every description.

"Corn" is to be understood in much the same way, or as a general term for cereal grain, though the appellation may at times be made specific. In Shakspere this word occurs upon at least twenty occasions, sometimes in reference to corn as an article of food; sometimes, as it has been for thousands of years (corn being an adjunct, as well as one of the factors of civilisation), as an emblem of prosperity and wealth—

No use of metal, corn, or wine, or oil.—*Tempest*, ii., 1.

I am right glad to catch this good occasion,
Most thoroughly to be winnowed, where my chaff
And corn shall fly asunder.—*Henry the Eighth*, v., 1.

Occasionally there is a fine image of some other kind, as in Cranmer's flattering prophecy of the grand reign of Elizabeth:—

> Her foes shake like a field of beaten corn,
> And hang their heads with sorrow.—*Ibid*, v., 4.

> Why droops my lord, like over-ripened corn
> Hanging the head at Ceres' plenteous load ?
> *2nd Henry the Sixth*, i., 2.

> First let me teach you how to knit again
> This scattered corn into one mutual sheaf.—*Titus*, v., 3.

When, in the *Midsummer Night's Dream*, i., 2, Oberon says,

> Am not I thy lord?

and Titania replies,

> Then I must be thy lady.   But I know
> When thou hast stolen away from fairy-land,
> And in the shape of Corin sat all day,
> Playing on pipes of corn, and versing love,

she refers particularly to oats, the stems of which, just before they begin to change colour, supply tolerably inflexible tubes, which, with a little trimming, can be converted into pipes, suitable in clever hands for imitation of such music as that of the "great god Pan," when he resorted to the "reeds by the river." In ancient times these little pipes were a favourite musical instrument with shepherds, as appears both from Ovid (*Met.*, i., 677) and from Virgil, whose reference to them is almost the first thing learned in school-boy Latin,

> Tityre, tu, patulæ recubans sub tegmine fagi,
> Silvestrem tenui Musam meditaris avenâ.—Ecl. i., 1, 2.

Mention of them is made also in Ecl. x., 51. Both authors make *avena*, when used in this connection, equivalent to *calamus* and *arundo*. The introduction of them in the *Midsummer Night's Dream* passage is peculiarly appropriate, the musician himself being so diminutive. They are spoken of also in *Love's Labour's Lost*, at the end:—

> When shepherds pipe on oaten straws,
> And merry larks are ploughmen's clocks.

As a grain good for horses, Shakspere refers to the OAT upon four or five occasions. BARLEY appears twice—*Tempest*, iv., 1; *Henry the Fifth*, iii., 5. RYE also twice:—

> Between the acres of the rye
>
> .   .   .   .
>
> Those pretty country folks would lie.
> *As You Like It*, v., 3.

> You sun-burned sicklemen, of August weary,
> Come hither from the furrow and be merry,
> Make holiday, your rye-straw hats put on.—*Tempest*, iv., 1.

WHEAT comes in rather more frequently; and now and then in very beautiful association:—

> O happy fair !
> Your eyes are lode-stars, and your tongue's sweet air
> More tuneable than lark to shepherd's ear,
> When wheat is green, when hawthorn buds appear.
> *Midsummer Night's Dream*, i., 1.

> As love between them like the palm might flourish;
> As peace should still her wheaten garland wear.
> *Hamlet*, v., 2.

The green fodder plants esteemed in the Elizabethan age

appear to have been precisely those which are valued by the farmer of to-day. CLOVER is mentioned in *Henry the Fifth*, v., 2; also, under the synonym of "honey-stalks," in *Titus Andronicus*, iv., 4. VETCHES are named in the *Tempest*, iv., 1; and HORSE-BEANS in the *Midsummer Night's Dream*, ii., 1. An exception must be made perhaps, as regards Burnet, named with clover in the *Henry the Fifth* passage, care for this plant as a cattle-food being now very limited. As one of Shakspere's plants, its place of course remains unchanged, and being frequent as a wild-flower in calcareous districts, to learn its pretty complexion is quite easy. Botanically, burnet is the *Poterium Sanguisorba*, the generic name referring to the pleasant flavour imparted by the leaves, which taste something like cucumber, to goblets of wine. Thus employed, says the old herbalist, "it yealdeth a certaine grace in the drinking," making the heart "merry and glad." The leaves, which come up in dense tufts, are very elegantly pinnate. The flowers are clustered in little spheres at the points of stems about nine inches high, the stamens hanging out in buff-coloured tassels, the pistils forming little crimson aspergilli.

SHAKSPERE'S WEEDS.

Weeds, so called, are wild-flowers in excess of number, and growing where the soil is wanted for the purposes of cultivation. Taken individually, few of the plants which are prone to trouble the gardener and the farmer through their fecundity and their perseverance, are inferior in

beauty of structure to the generality of the garden
favourites. Isolated by the wayside, and in the pride of
their summer, they seldom fail to please. Many of the
larger kinds no doubt, for want of training, become
ragged and disorderly, but even then they redeem themselves by their lustre. Nothing in the costliest garden is
more splendid than a full-grown cotton-thistle, a scarlet
corn-poppy, or a well-developed specimen of the azure
bugloss, *Echium vulgare.* At any time, it is not so much
what the plant is, as *where* it is, that constitutes either
the gem or the weed; just as, upon the converse, in
regard to the affections, it is not so much in *places* that
repose is found, as in *persons.* Shakspere shows us many
a time that weeds, although impedimenta, can still be
made elements of most beautiful pictures. Nowhere in
his writings is there a more perfectly finished though
sorrowful description than that one of the calamities
induced by war, under the figure of weeds, in *Henry the
Fifth,* briefly referred to when speaking of the cowslip:—

> My duty to you both, an equal love,
> Great kings of France and England !
>   That I have laboured
> With all my wits, my pains, and strong endeavours
> To bring your most imperial majesties
> Unto this bar and royal interview,
> Your mightiness on both parts best can witness.
> Since, then, my office hath so far prevailed,
> That, face to face, and royal eye to eye,
> You have congreeted; let it not disgrace me,
> If I demand, before this royal view,
> What rub, or what impediment there is,

Why that the naked, poor, and mangled peace,
Dear nurse of plenties, arts, and joyful births,
Should not, in this best garden of the world,
Our fertile France, put up her lovely visage ?
Alas ! she hath from France too long been chased,
And all her husbandry doth lie on heaps,
Corrupting in its own fertility.
Her vine, the merry cheerer of the heart,
Unprunèd dies; her hedges even-pleached,
Like prisoners wildly overgrown with hair,
Put forth disorder'd twigs : her fallow leas,
The darnel, hemlock, and rank fumitory,
Doth root upon; while that the coulter rusts,
That should deracinate such savagery.
The even mead, that erst brought sweetly forth
The freckled cowslip, burnet, and green clover,
Wanting the scythe, all uncorrected, rank,
Conceives by idleness, and nothing teems
But hateful docks, rough thistles, kecksies, burs,
Losing both beauty and utility.
And, as our vineyards, fallows, meads, and hedges,
Defective in their natures, grow to wildness,
Even so our houses, and ourselves, and children,
Have lost, or do not learn, for want of time
The sciences that should become our country,
But grow, like savages,—as soldiers will,
That nothing do but meditate on blood,—
To swearing, and stern looks, diffused attire,
And everything that seems unnatural ;
Which to reduce into our former favour,
You are assembled; and my speech entreats
That I may know the let, why gentle peace
Should not expel these inconveniences,
And bless us with her former qualities.—(v., 2.)

Mark, before going into the consideration of the
individual weeds, the powerful reminder supplied in this

fine passage, that Nature, although from one point of
view, inanimate and passionless, is capable of setting
forth both joy and sorrow. All the highest poetry
recognises in Nature a life higher than the simply
physiological. Neglected, it runs to waste. Carefully
tended; fed with human love, it rejoices. Mark, also,
the Shaksperean association of weeds with what is
dishonourable in life as well as with the lamentable.
To the bridal couch of virtue he gives flowers:—

> QUEEN: Sweets to the sweet: Farewell,
> I hoped thou should'st have been my Hamlet's wife.
> I thought thy bride-bed to have decked, sweet maid,
> And not have strewed thy grave.—*Hamlet*, v., I.

On the contrary,—

> No sweet aspersion shall the heavens let fall
> To make this contract grow; but barren hate,
> Sour-eyed disdain, and discord shall bestrew
> The union of your bed with weeds so loathly
> That you shall hate it both.—*Tempest*, iv., I.

"Aspersion" is one of the words which illustrate how
curiously, with lapse of time, the original or literal sense
may disappear, or nearly so, the figurative or metaphorical
one alone remaining. The literal sense of "aspersion,"
that in which Shakspere employs the word, is a scattering or sprinkling, as of rain-drops from the clouds, and
when the "holy water" is cast upon the people during
the ceremonies of the Catholic Church. At present it
is seldom heard except in the extended sense of scattering calumny or false accusations. Three curious old
terms in the *Henry the Fifth* passage also deserve notice,

"deracinate," "teem," and "the let." To deracinate is literally to tear up by the roots; as again in *Troilus and Cressida*, i., 3:—

> Frights, changes, horrors,
> Divert and crack, rend, and deracinate
> The unity and calm of married states
> Quite from their fixture.

"Teem" is the Anglo-Saxon *teman*, to bring forth, as in the miserable king's imprecations upon his shameless daughter:—

> Hear, nature, hear;
> Dear goddess, hear ! Suspend thy purpose, if
> Thou didst intend to make this creature fruitful!
> Unto her womb convey sterility !
> Dry up in her the organs of increase,
> And from her derogate body never spring
> A babe to honour her ! If she must teem,
> Create her child of spleen, that it may live,
> And be a thwart disnatur'd torment to her!—*Lear*, i., 4.

A "let" signified originally a hindrance, as in *Hamlet*, "I'll make a ghost of him that lets me." Shakspere's contemporaries use the word in exactly the same sense:—

> It had been done ere this, had I been counsel:
> We had had no stop, no let.—*Ben Jonson.*

In matters of this kind, he is usually his own interpreter. The number of words which occur but once in the dramas must be relatively very small.

Several of the weed-names in the *Henry the Fifth* passage must be understood as collective. The names of individual kinds of cornfield interlopers have in all ages been employed as appellations for the aggregate;

and though science may find it convenient to restrict
the application, as long as the world lasts, the *usus
loquendi* will take its ancient course. Shakspere probably
meant nothing definite by "darnel," though the botanists
tie down this name to the *Lolium temulentum.* They do
so because tradition says that this grass was the *infelix
lolium* of Virgil, which is the same as identifying it with
the αἶρα of the Greeks. The sense is precisely that of
the ζιζάνιον of the parable of the sower, in the A.V.
rendered "tares." Fuchsius, in 1542, folio 127, figures
the purple-flowered corn-cockle, Agrostemma, or *Githago
segetum,* under the name of lolium, and, as if in translation of Fuchsius, "darnel" is the name given to this
identical plant by Shakspere's contemporary, Drayton:—

> The crimson darnel-flower, the blue-bottle, and gold,*
> Which though esteemed but weeds, yet for their dainty hues,
> And for their scent not ill, they for their purpose choose.

Cockle and lolium seem, in truth, to have meant, in the
Shaksperean times, very much the same thing. Chaucer
had already used "cockle" for cornfield-weeds in general.
Old Latimer, speaking of Satan, in his famous Sermons,
p. 71, asks vehemently "who is able to tell his diligent
preaching which every day and every hour laboureth to
sow cockle and darnel?" Newton, in the *Herbal to the
Bible,* 1587, observes: "Under these names of cockle
and darnel are generally understood not themselves only,
but also all harmefull encumbrances, lets, hurts, and

---

* "Golds," in the Elizabethan times, were the corn-marigold,
Chrysanthemum segetum. Spenser calls them "goolds."

annoyances, which doe any way hinder the growth of
corne." "Darnel" appears again in *King Lear*, iv., 4,
when the wretched old monarch, his reason departed,
is met

> As mad as the vexed sea; singing aloud;
> Crowned with rank fumiter and furrow-weeds,
> With hardocks, hemlock, nettles, cuckoo-flowers,
> Darnel, and all the idle weeds that grow
> In our sustaining corn.

"Cockle," anciently "kokylle"—a name derived from
caucalis, the Greek καυκαλὶς, the wild carrot—is mentioned in *Coriolanus*, iii., 1, in a figurative sense,—

> We nourish 'gainst our senate
> The cockle of rebellion,—

and again in *Love's Labour's Lost*, iv., 3, where it seems
to preserve some kind of proverbial phrase,—

> Allons! allons! Sowed cockle reaped no corn.

The sense is here that, beginning with deceit, the end
can be only emptiness and disappointment, as verified by
the lines which immediately succeed:—

> And justice always whirls in equal measure;
> Light wenches may prove plagues to men forsworn;
> If so, our coffer buys no better treasure.

For convenience' sake, as with "darnel," it may be
allowed to limit Shakspere's "cockle" to the Githago,
one of the handsomest plants of its tribe, the purple
flowers an inch across, and lifted to the level, nearly,
of the ripening ears.

So with "Hemlock." The derivation of this name
shows it to have been originally applied to large and

vigorous members of the natural order Umbelliferæ in
general, both poisonous and innocuous, as indeed it is to
this day with the non-discriminating. Foremost among
these would be the genuine hemlock of Botany, *Conium
maculatum*, the cow-bane, *Cicuta virosa*, and the *Angelica
sylvestris*. Nothing more definite can be asserted in
regard either to the *Henry the Fifth* passage, or to the
other in *King Lear*. Shakspere may quite possibly have
adopted the word from the Old Testament, where in two
places (Hosea, x., 4, and Amos, vi., 12), it is put for the
Hebrew *rō'sh*, a term of unknown signification, except as
denoting some kind of corn-field intruder. It is these
self-same plants, in any case, the larger Umbelliferæ,
which are to be understood by the term "kecksies," or
kex, applied to them by metonymy, kecksies being the
stiff, round, tubular stems, useful, when stripped of the
leaves and trimmed, for various domestic purposes. The
name may be connected very specially with the stems of
the angelica, that beautiful species of its great family
which haunts the water-side in every stream-fed meadow,
reserves its large round delicately lilac umbel till the
arrival of autumn, and is always so reluctant to go.

The name occurs a third time in *Macbeth*, when the
witches are preparing their horrible broth,—

> Root of hemlock, digg'd in the dark.

Here we may legitimately understand allusion to the
genuine conium, the deadly "hemlock" *par excellence*,
"whosoever taketh which into his body," says Lyte,

"dieth remedilesse." "Root" does not necessarily mean merely the *radix*, being put, again by metonymy, for the plant as a whole, a very common usage, and well illustrated in another passage in the same play,—that one where Banquo, after the interview with the infernal trio, says to Macbeth,—

> Were such things here, as we do speak about?
> Or have we eaten of the insane root,
> That takes the reason prisoner?—(i., 3.)

The root or plant intended here would seem to be henbane, *Hyoscyamus niger*, described as far back as the time of Dioscorides, as causing madness, an idea repeated in various writers of the Shaksperean age, as Boord, in the *Brevyary of Health*, 1557, and Bulleyn, who says, "if it be used either in sallet or in potage, then doth it bring frenzie, and whoso useth more than four leaves shall be in danger to sleepe without waking."

By "Hardocks," mentioned in the Lear passages, are probably intended those roughest and coarsest of our native plants, the more-usually-called "burdocks," in their one or two kinds, the *Arctium Lappa* and *Arctium Bardana* of the botanists. Haunting farm-land, although the foliage is liked by artists for their foregrounds, they are almost too cumbersome to be admired. They possess a certain curious interest, nevertheless, in the abundance of their spherical heads, purple-topped while young, and covered at all times with little hooks, which catch one's clothing, or anything else that is soft, on the instant, and never willingly let go. Hence the familiar

name of "burs," given to the plant itself, for brevity'
sake, in *Henry the Fifth*, and so aptly employed in
metaphor, upon no fewer than four separate occasions:—

> Nay, friar, I am a kind of bur, I shall stick.
>> *Measure for Measure*, iv., 3.

> Our kindred, though they be long ere they are wooed, are con-
> stant, being won; they are burs, I can tell you.—*Troilus and
> Cressida*, iii., 2.

Then in the *Midsummer Night's Dream*, iii., 2, and
lastly, when Rosalind and Celia, before setting off for
the woods, are chatting in somewhat melancholy mood;
Rosalind herself, even, dejected for the time, for her
father's sorrows are weighing upon her mind:—"O how
full of briars is this work-a-day world!" Happy those
who in such cases have a loving Celia at hand to reply,
"They are but burs, cousin, thrown upon thee in holiday
foolery. If we walk not in the trodden path, our very
petticoats will catch them." But the smile cannot
come just yet—"I could shake them off my coat; these
burs are in my heart." Were the occasion more profoundly serious, the words would remind one of the iron
that "entereth into the soul."

Shakspere's "Thistles" may be thought of as well
represented in the stalwart and universal *Carduus lan-
ceolatus*, the "spear-thistle" of the vernacular, which will
serve alike for the intruding weed, and for the beautiful
picture in the *Midsummer Night's Dream*, of the bee
sucking the honey from the well-charged crimson bloom.
Shakspere, who had the observing eye, that inestimable

privilege, could not fail, however, to have often noticed
the wand-like stems of the *palustris*, and their crowded
clusters; the *arvensis*, that returns to the charge again
and again, with valour so troublesome; and that
supremely handsome plant of the green margins of
little-travelled roads, and of airy green downs, the
nodding musk-thistle, *Carduus nutans*, so fragrant, but
alas, so vicious to the fingers that would molest it.

Fairly precise as the application of this name is in
Shakspere, "thistle" must not be accepted as such
in literature generally, and, least of all, as regards its
employment in the A.V. of the Old Testament, where
it is used in the broad and general sense of prickly corn-
field plants.   In the famous Genesis passage, iii., 18,
the Hebrew word, *dardar*, if it means anything specific,
denotes, perhaps, the *Tribulus terrestris*, or "caltrops;"
and the same will then be meant in Hosea, x., 8.   The
other Hebrew word translated "thistle," *chōach*, is so
vague in the literal sense, that it is rendered indifferently,
thistle, thorn, and bramble.

Shakspere's "dock," *Tempest*, ii., 1, may safely be
understood as the large, coarse *Rumex obtusifolius* of
the wayside.   The "mallows" of the same line are the
Malva sylvestris, a plant incapable of keeping itself tidy,
but with beautiful purple flowers.   "Knot-grass," *Midsummer Night's Dream*, iii., 2, is the *Polygonum
aviculare*, called "hindering," because of a droll superstition that if a boy drank an infusion of the stems and
leaves, it would arrest his growth.   "Fumitory" is that

light and pretty little annual, with purple and rosy
flowers, which botanists call *Fumaria officinalis*, and the
name of which is undividedly its own. The furrows
retain it long after the corn and barley have been
garnered. It is one of the sweet little lingering things
which, though Autumn gives daily warning, are always
so loth to depart, and is never prettier than when
"smiling through its tears" in the dewy beams of
an October sunrise. "Rank" fumiter means simply
"luxuriant," the epithet having here the same sense as
in Genesis xli., 5.

Many of the plants which vex the gardener by their
intrusion where not wanted, really and truly deserve the
epithet of rank, or luxuriant to a degree that renders
them harmful. These are referred to in *Richard the
Third*, ii., 4:—

> Grandam, one night, as we did sit at supper,
> My uncle Rivers talked how I did grow
> More than my brother. "Ay," quoth my uncle Gloster,
> "Small weeds have grace; great herbs do grow apace;"
> And since, methinks, I would not grow so fast,
> Because sweet flowers are slow, and weeds make haste.

Of Shakspere's agricultural weeds there remains only
the Nettle, the familiar stinging plant from which
Ophelia's is differentiated alike by its beautiful golden
flowers and inability to hurt, the bloom of the *Urtica*
being green and insignificant. The latter is mentioned
by Shakspere upon eleven distinct occasions, several of
which involve metaphor, as most natural, considering the

Q

temper of the plant. One reference alone calls for special notice, and this one chiefly because of the proposed curious change of the reading. In *Twelfth Night*, ii., 5, when keen-witted, sarcastic Maria—Olivia's maid—comes tripping into the garden, "Here," exclaims Sir Toby, with his accustomed drollery, "Here comes the little villain: How now, my nettle of India?" Some of the critics say this ought to be "mettle." Others would have "metal," understanding the phrase to mean "gold of India," and to be equivalent to "my darling." There is no necessity for any such change. In the scene closely preceding (ii., 3), Sir Toby has already addressed the "little villain" as Penthesilea, comparing her, by the use of that name, to the famous queen of the Amazons, killed in single combat with Achilles, a fate declaring her spirit and courage. In i., 3, where her behaviour to the pair of old knights shows very plainly what she is made of, Sir Andrew calls her "fair shrew." Now, Maria all over, she is about to cheat her "trout" Malvolio. Sir Toby perceives exactly what is ahead. He aptly compares her to the lively weed, and "of India" comes in adjectively as a metaphor of the warmth of purpose with which she will operate upon her victim. The idea comes up again in *3rd Henry the Sixth*, iii., 3:—

> Nay, mark, how Lewis stamps as he were nettled;
> I hope all's for the best.

Also in *1st Henry the Fourth*, i., 3:—

> Why, look you, I am whipped and scourged with rods,

Nettled, and stung with pismires, when I hear
Of this vile politician, Bolingbroke.

It may not be superfluous, after all, to quote
the nettle passage in *Othello*, since it illustrates
in the same manner as when Apemantus appeals
to Timon, "Shame not these woods," — how profound was Shakspere's knowledge of the possibilities
of human nature. Iago has the heart and the desires
of a demon. Shakspere allows him, nevertheless,
to give lessons in ethics. "Virtue? a fig! 'Tis in
ourselves that we are thus, or thus. Our bodies
are our gardens, to the which our wills are gardeners;
so that if we plant nettles, or sow lettuce,—set hyssop,
and weed up thyme,—supply it with one gender of
herbs, or distract it with many,—either to have it sterile
with idleness, or manured with industry;—why the power
and corrigible authority of this lies in our wills" (i., 3).
The passage is of course to be understood as entirely
metaphorical. By lettuce, hyssop, and thyme, he means
the agreeable possessions and pleasures of the heart which
reward discreet self-culture. Nettles, on the other hand,
represent the miserable, discontented hours, the suscep-
tibilities to mean mortification, which inevitably arise
upon neglect of it, with those most despicable of all
things, envy and selfishness.

"Weed," Anglo-Saxon *wæd*, meaning clothes or apparel,
a sense of the word now almost forgotten, except in
"widow's weeds," also occurs several times in Shakspere.
In the *Winter's Tale*, iv., 3, when Perdita is dressed

somewhat showily for the reception of her father's guests, Florizel says:—

> These your unusual weeds to each part of you
> Do give a life; no shepherdess, but Flora,
> Peering in April's front. This your sheep-shearing
> Is as a meeting of the petty gods,
> And you the queen on't.

So in the *Midsummer Night's Dream*, ii., 2:—

> There sleeps Titania, some time of the night,
> Lull'd in these flowers with dances and delight;
> And there the snake throws her enamell'd skin
> Weed wide enough to wrap a fairy in.

The grey transparent beauty of the cast-off skin of the snake is well known. The amplitude is more than enough: the variegated delicacy of the natural pattern might supply hints for the fabricators of mantles other than such as fairies need.

Chapter Twelfth.

THE WILDERNESS AND THE WAYSIDE.

Love adds a precious seeing to the eye;
A lover's gaze will gaze an eagle blind;
A lover's ear will hear the lowest sound.
Love's Labour's Lost, iv., 3.

ROM the neglected farm to the wilderness is but a step; the productions are, nevertheless, in many respects quite different. The ligneous plants found in the hedgerows, where all is, no doubt, as well as in the wilderness, comfortable, are themselves always suggestive of the untamed. The hedgerows, in truth, constitute a kind of charming border-land between scenes of exact culture and the desert. They contain little that has been deliberately planted; in parts they are altogether barbarian; yet it quite as often happens that their productions are no less estimable

for economic use than those of the meadow and the cottage garden. Their happy intermediate character is declared also by the graceful decorative plants which often intermingle plentifully with the ruder ones—the vine-like bryony, the glossy tamus, the pencilled wood-vetch, the scarlet and orange-berried spindle-tree—not to mention the wild clematis, the twice-blooming honey-suckle, and the dog-roses that never satiate. Sheltered by these larger plants, on the slopes below them grow wild-flowers that enjoy life upon the hedge-bank more than in any other habitat which suits their nature, as the germander-speedwell and the snowy mollugo; and no slight matter is it for these, that along with all other wilderness and semi-wilderness wild-flowers, they have the privilege of immunity from the swift ruin wrought by the sickle and the scythe.

Shakspere refers to at least a dozen of the hedgerow and wilderness class of our indigenous Flora, and, as a rule, the allusions, if not distinctly pleasing and poetical, have nothing at all about them that offends. Con-spicuous among the fruit-bearing kinds, we have, in the first instance, the wild apple, often called simply the "crab," as in a passage that would be beautiful in every sense were it not spoken under the influence of wine:—

> I'll show thee the best springs; I'll pluck thee berries;
> I'll fish for thee, and get thee wood enough.
>
> .   .   .   .   .   .
>
> I pr'ythee let me bring thee where crabs grow;
> And I with my long nails will dig thee pig-nuts;
> Show thee a jay's nest, and instruct thee how

To snare the nimble marmazet. I'll bring thee
To clustering filberds, and sometimes I'll get thee
Young sea-mells from the rock.—*Tempest*, ii., 2.

The harshness and the acidity of the fruit are proverbial:

PETRUCHIO: Nay, come, Kate, come, you must not look so sour.
KATHERINE: It is my fashion when I see a crab.
PETRUCHIO: Why here's no crab, and therefore look not sour.
Taming of the Shrew, ii., 1.

It was esteemed, nevertheless, in rustic fireside merry-makings,—

When roasted crabs hiss in the bowl,
Then nightly sings the staring owl.
Love's Labour's Lost, v., 2.

PUCK: And sometimes lurk I in a gossip's bowl,
In very likeness of a roasted crab.
Midsummer Night's Dream, ii., 1.

When pressed, as if for cider-making, these wild apples yielded the thin and rather poor condiment for those who could not afford a better one of the kind, called verjuice. In the Elizabethan age it was in common use, and though not mentioned by Shakspere, seems to be pointed to in the name of the silly sub-constable Verges, in *Much Ado;* just as foolish, self-complacent Dogberry, his superior, seems to have got his name from the worthless fruit of the dogwood, *Cornus sanguinea.**

---

* This must not be confounded with the cherry-like fruit of the *Cornus mascula*, celebrated by the old Roman poets. The berries of the *sanguinea*, very common in the south of England, are small, black, and bitter.

The wood of the crab-tree was valued for the sake of its toughness:—

> Fetch me a dozen crab-tree staves, and strong ones.
> *Henry the Eighth*, v., 3.

The Hazel-nut. After the crab comes the hazel-nut—only the produce, never the tree itself, except laterally, in the first portion of Petruchio's gallant comparison,—

> Kate, like the hazel-twig,
> Is straight and slender; and as brown in hue
> As hazel-nuts, and sweeter than the kernels.
> *Taming of the Shrew*, ii., 1.

Are we to to think of Baptista's shrewish daughter as gipsy-faced? Was she another "Nut-brown mayde," that faithful creature to whom good and evil were alike indifferent, so that she could share the fate of the man she loved, even in the wilderness?—

> Among the wylde dere, such an archere
> As men say that ye be,
> May ye not fayle of good vitayle,
> Where is so great plentë.
> And water clere of the ryvere
> Shal be full swete to me,
> With which in hele, I shall ryght wele
> Endure, as ye shall see.
> And, or we go, a bedde or two,
> I can provyde anone,
> For, in my mynde, of all mankynde
> I love but you alone.

Perhaps he intended that her *eyes* were of that beautiful colour, likening them to the smooth brown shell after the manner of Mercutio in *Romeo and Juliet:*—"Thou wilt quarrel with a man for cracking nuts, having no other reason but because thou hast hazel eyes."

Shakspere knew full well the ways, the manners and customs, of simple woodland creatures; his zoology is often as interesting as his botany:—

> TITANIA: I have a venturous fairy that shall seek
> The squirrel's hoard, and fetch thee new nuts.
> *Midsummer Night's Dream*, iv., 1.

And how beautiful the introduction of the little animal as a skilled artificer !

> O then I see Queen Mab hath been with you !
> She is the fairies' midwife, and she comes
> In shape no bigger than an agate-stone
> On the forefinger of an alderman, .
> Drawn with a team of little atomies,
> Athwart men's noses as they lie asleep.
>
> Her chariot is an empty hazel-nut,
> Made by the joiner squirrel or old grub,
> Time out of mind the fairies' coachmakers.
> *Romeo and Juliet*, iv., 1.

Whatever people may think of Folk-lore in general—a natural outcome of the human fancy, therefore wisdom to the wise, though folly to the incurious, it is impossible to resist capture by Queen Mab, the more particularly since it is in Shakspere that she first appears in the place assigned to her. No earlier instance, that is to say, occurs in literature of her being styled the Queen of the Fairies. Gloriana, the "Faery Queene" of Shakspere's noble contemporary, Edmund Spenser, it hardly needs the saying, is not queen of the fairies at all, but of

chivalry and the knightly virtues. The history of the
royal lady is full of charming interest. Enough for the
present to say that while in powers and attributes Queen
Mab is another rendering of the idea pourtrayed in
classical Diana, the name itself is Keltic, and that she
is the Méave or Mebdh of ancient Irish story. Folk-lore,
woven in so many hues, pleased Shakspere most upon
its fairy side. How often the little beings are allowed
to step to the front we have seen already. They compensate the total absence from his dramas of little girls.

Five or six other allusions to nuts and nutshells might
be quoted, but with one exception they carry little of
importance. This is in *All's Well*, when Lafeu says
of the foppish and vacant Parolles, in well-chosen
metaphor, "There can be no kernel in this light nut:
the soul of this man is his clothes: trust him not in
matter of heavy consequence" (ii., 4). In the *Tempest*
nuts are called "filberts," the name, strictly speaking,
of the garden or cultivated form of the fruit, but that
the hazel-nut is meant is shown by the context.

The Blackberry is mentioned thrice;—in *Troilus and
Cressida*, v., 4, as an emblem of worthlessness; in *1st
Henry the Fourth*, Falstaff the speaker, as an emblem of
abundance—"Give you a reason on compulsion? If
reasons were as plenty as blackberries, I would give
no man a reason on compulsion;" and further on, in
the same scene, again by Falstaff, while joking with
Prince Henry—"Shall the blessed son of heaven prove a
micher, and eat blackberries?" The allusion is here to

boys who, playing truant from school, and afraid to go home, to escape observation wander up and down the lanes and by hedgerows, to "mich" signifying to lurk out of sight, or hide one's self. Of all our wild or native fruits, this one is without question the most interesting, bound up as it is with the history of childhood, joy and grief going hand in hand over the gathering; commemorated, moreover, in many a simple old verse :—

> Their pretty lips with blackberries
> Were all besmeared and dy'd,
> And when they saw the darksome night,
> They sat them down and cry'd.
> > *The Children in the Wood.*— *Roxburghe*
> > *Ballads,* ii., 220.

The fruit may, perhaps, have been what Caliban promised to find,—

> I'll pluck thee berries,—

though it can hardly be the same which he tells us Prospero gave him in the beginning of his servitude :—

> When thou cam'st first,
> Thou strok'st me, and mad'st much of me; would'st give me
> Water with berries in it;
> > And then I loved thee,
> And show'd thee all the qualities of the isle,
> Cursèd be I that did so !

What Shakspere intends in the latter passage by "berries" is an open question, met, so far, by no satisfactory conjecture. It is probable, however, that he really means blackberries when in the *Midsummer Night's Dream,*

iii., 1, he mentions dewberries. Titania, giving her
pleasant behest to the four fairies, says:—

> Be kind and courteous to this gentleman;
> Hop in his walks, and gambol in his eyes;
> Feed him with apricocks and dewberries,
> With purple grapes, green figs, and mulberries;
> The honey-bags steal from the humble-bees,
> And for night-tapers, crop their waxen thighs,
> And light them at the fiery glow-worm's eyes,
> To have my love to bed.

For, compared with the fruit of the *Rubus fruticosus*,
that of the *cæsius* is insipid; it is never collected for
eating, and certainly would not be such as Titania
would associate, meaning it to be a treat, with figs and
apricots. In Shakspere's time, moreover, the two names
were used indifferently. Lyte, p. 661, describing the
cæsius, says it "is called a dewberrie or blackberrie."
Possibly, in the present instance, "dewberry" may have
been introduced as more euphonious. Shakspere could
hardly fail to be acquainted with it botanically, the large
round drupeolæ being covered with a peculiar glaucous
bloom, different altogether from the rich and jetty lustre
of the blackberry *ipsissima*.

The plant producing the genuine blackberry is mentioned many times, either as the "bramble," or inclusively,
in the large company of the "thorns." The former name
occurs in *Venus and Adonis*, and in *As You Like It*,
"There is a man hangs odes upon hawthorns, and elegies
upon brambles" (iii., 2). Thorn, meaning the blackberry,
appears in the *Midsummer Night's Dream*, iii, 2,—

For briers and thorns at their apparel snatch;

also in *3rd Henry the Sixth*, iii., 2,—

And I, like one lost in a thorny wood.

By "thorns," however, in several other places, we are to understand not so much brambles, or Rubi, as prickly and spinous plants which grow underfoot, after the manner of the rest-harrow, *Ononis spinosa*, though there is nothing in any passage upon which to base an exact identification, even were this important. "Thorns and thistles" present a sufficiently exact and vivid picture of the plants said to be contemporaneous with the moral declension of mankind:—

The care you have of us,
To mow down thorns that would annoy our foot,
Is worthy praise.—*2nd King Henry the Sixth*, iii., 1.

Another prickly wilderness plant appears in a line made famous by the amount of dispute it has given rise to. Shipwreck in progress—"we split! we split!" and drowning imminent—"Now," says good old Gonzalo, "would I give a thousand furlongs of sea for an acre of barren ground, long heath, brown furze, anything. The wills above be done, but I would fain die a dry death" (*Tempest*, i., 1). The line is thus printed by all modern editors in deference to the first folio, 1623, which is, *literatim*, "long heath, browne firrs." It was suggested, however, by the "Oxford Editor," Sir Thos. Hanmer, in 1744, that the true reading would in all probability be, "ling, heath, broom, furze—anything,"—anything, that is to say, which means land, and not water, so that we do not drown. Hanmer, Dr. Johnson tells us, possessed that first and most essential requisite of the genuine critic, "the intuition by which the poet's intention is immediately discerned." He comes, accordingly, with excellent credentials, and that his judgment has plenty to sustain it, is undeniable. "Ling" is a well-known provincial name for the beautiful wilderness undershrub, with innumerable glossy lilac flowers, commonly called heather, the *Calluna vulgaris.* It is very different from the genuine heaths, or Ericas, so that there is no tautology in the introduction of the word, and Shakspere had very probably seen the plants in company, since the Calluna is a frequent associate of the no less lovely crimson heath, *Erica cinerea,* which if "heath" means anything at all, it will mean *here.* "Broom," assuming Sir Thomas Hanmer's conjecture to be well founded, is the well-known shrub, rendered conspicuous in early summer by its countless dishevelled golden butterflies. Furze, the never bloomless, the flowers of rival gold, and exhaling the odour of cocoanut, grows in our country more profusely even than the Spartium. Is it possible that the truth lies between the two extremes, and that the reading should be, "ling, heath, brown furze?" The great objection, both to "long" and to "brown," is that in an emergency so terrible Gonzalo would not be likely to employ epithets. He would just run over the names of the plants, and leave adjectives till a happier time. If Sir Thomas Hanmer's interpretation be accepted as the most reasonable one, the line in question, instead of adding only two plants to the general catalogue, heath and furze, will thus really make an addition of four.

There is something better for us in the connection than the minute botany,—the character of the "honest old counsellor of Naples" himself, the only man on board who in those terrible moments preserved his cheerfulness, and nourished his heart with hope—a mere coincidence perhaps, but in any case we cannot forget that his previous life had been one of virtue. It was Gonzalo who, when Prospero was driven from his home and country, stood the thoughtful "friend in need." Miranda, whose innocent prattle and sweet child-smiles had sustained her father during their first sad years upon the desolate island, "infused with a fortitude from heaven,"—a great light may shine through a very small window,—now that she is old enough to understand, is told the story of their wrongs. And how, says she, " came we ashore?"

> PROSPERO: By Providence divine.
>   Some food we had, and some fresh water, that
>   A noble Neapolitan, Gonzalo,
>   Out of his charity (who being then appointed
>   Master of this design) did give us; with
>   Rich garments, linens, stuffs, and necessaries
>   Which since have steadied much. So of his gentleness,
>   Knowing I loved my books, he furnished me,
>   From my own library, with volumes that
>   I prize above my dukedom.—*Tempest*, i., 2.

Later on in the play, furze turns out to be nearer than Gonzalo imagined. Once upon shore, his wish would

have been gratified without going very far. Prospero,
having given mischief-loving little Ariel charge of Caliban
and his fellow-rascals for punishment, now inquires what
has he done,—

> Where did'st thou leave these varlets?
> ARIEL: I told you, sir, they were red-hot with drinking.
> Then I beat my tabor,
> At which, like unback'd colts, they pricked their ears,
> Advanced their eyelids, lifted up their noses,
> As they smelt music. So I charmed their ears
> That, calf-like, they my lowing followed, through
> Toothed briers, sharp furzes, pricking goss, and thorns,
> Which entered their frail shins. At last I left them
> I'the filthy mantled pool beyond your cell,
> There dancing up to the chins!

What a subject for a comic picture! As yet, our artists
have hardly touched the edge of Shakspere, though
there is material enough in him for galleries.

Ariel's "pricking goss" is usually understood to be one
of the smaller kinds of furze, and some have connected
the name, definitely, with the *Genista Anglica*. But there
is no justification for this. The word is but a varied form
of the low Latin *gorra* or *gorassi*, denoting brushwood
in general. It is only in modern times that gorse has
been made a synonym of furze, and Shakspere in using
goss had certainly nothing specific in view. Gorassi
when cut down and dried became " bavin," the coarse
fuel mentioned in *1st Henry the Fourth*, iii , 2, in capital
metaphorical allusion to the amusement afforded by
poor inane buffoons, easily got out of them, over in a
moment:—

> He ambled up and down,
> With shallow jesters, and rash bavin wits,
> Soon kindled, and soon burned.

Rough and tangled wilderness shrubs, with plenty of the tall coarse fern commonly called "bracken," *Pteris Aquilina*, would constitute, as to-day, the Shaksperean "brake," such a one, for example, as when Demetrius shows how cruel a man can be to a woman who for her own part loves fondly as life:—

> HELENA: Your virtue is my privilege for that.
> It is not night when I do see your face,
> Therefore I think I am not in the night;
> Nor does this wood lack worlds of company,
> For you, in my respect, are all the world.
> Then how can it be said, I am alone,
> When all the world is here to look on me?
> DEMETRIUS; I'll run from thee, and hide me in the brakes,
> And leave thee to the mercy of wild beasts!
>
> *Midsummer Night's Dream*, ii., 2.

The Aquilina, which often attains, in the wilderness and in woods, the stature of seven or eight feet, would also be the principal feature of the sylvan covert upon the margin of the nibbled turf, where the keepers of the chase, armed with their crossbows, waited the coming of the deer:—

> Under this thick-grown brake we'll shroud ourselves,
> For through this laund anon the deer will come;
> And in this covert will we make our stand.
>
> *3rd King Henry the Sixth*, iii., 1.

That the plant was named, like "heath," from its habitat, and not the habitat from the plant, has been plainly

R

certified by the etymologists. It is quite legitimate,
nevertheless, in any such picture, when deer are mentioned, to at once fill the "brake," mentally, with the
fern, since there is nothing as tall as themselves in which
hart and roebuck take more delight. Bracken and the
dappled deer, when nature is allowed her own free way,
always imply one another. Whenever on the frontiers of
some sweet green undulating down, where "the heavens'
breath smells wooingly," a little forest of Aquilina may
be descried, be sure the pretty creatures are not far off.

No other allusion to ferns, as growing plants, occurs
in Shakspere. Even this one to the bracken (although
Lyte uses "brake" as a synonym of "female fern") can
be claimed only as by implication. He mentions the
"seed," the brown spores cast from the spangles upon
the under-surface—this, however, not as a botanical
object, but simply in reference to one of the celebrated
old mediæval superstitions. Being of impalpable minute-
ness, itself almost invisible, to be dusted over with fern-
seed, according to the doctrine of "Signatures," gave
absolute invisibility,—

We have the receipt of fern-seed; we walk invisible,

1st King Henry the Fourth, ii., 1.

The "receipt," so called, was the formula of magic
words to be employed during the process. Allusions to
the superstition are often met with in contemporaneous
literature, but never once do we come upon certificate
that the belief is sound! Old Turner (1562) speaks of
collecting it from the fronds of the shield-fern, comparing

the slaty-purple lids, or indusia, before they have begun to curl, to poppy-seeds. Seed, he adds, hath also been found upon "brakes," on "midsomer even." The sporiferous braid which runs round every leaflet of the pteris is not in perfection till several weeks later in the season, so we must conclude that there is here a little infusion of the fable, since the efficacy of the magic dust came largely of its being gathered during those lovely hours which offer a hand alike to sunset and to Aurora.

The Ivy.—Never, in the sylvan wilderness, do the eyes rest upon a spectacle more delightful than that of the aspiring "ivy green," as it creeps up trunk and branch, till, overweighted with its own wealth, it rolls over in those sumptuous and swelling masses, the very perfection of perennial *chiar'oscuro*, which bear the golden spheres of honeyed bloom, and seem emblems of immortality. How beautiful, too, that younger life set forth in the close embrace of little shoots that sport in all sweet shades of amber and ruddy bronze. Shakspere mentions ivy upon three occasions. In the *Midsummer Night's Dream*, iv., 1, it is associated with the honeysuckle of the woodlands, as an emblem of tender affection:—

> TITANIA: Sleep thou, and I will wind thee in my arms.
> Fairies, begone, and be all ways away.
> So doth the woodbine—the sweet honeysuckle—
> Gently entwist—the female ivy so
> Enrings the barky fingers of the elm.

"Female" is an epithet adopted from the early botanical

writers, who, imitating the herbalists of ancient Greece, bestowed it upon the flower and fruit-bearing individuals of many different species of plants in which there is no real distinction of sex, but which are often individually barren, or given exclusively to the production of leaves. Plenty of plants exist, many of them wild in England, which are really and truly illustrative of difference of sex. With these, however, ivy does not count, the structure being no less consummate than that of the lily and the rose. So long as the idea prevailed that leaves without flowers implied masculine, as distinguished from feminine nature, ivy might well be regarded as male and female—for it is the custom of this ambitious plant never to produce flowers and berries while ascending. However lofty the vertical surface up which it runs, on it goes, onwards, leafy, but still quite sterile. Not until disengaged, and rolling as it were into space, does the inflorescence come forth, and then it is developed, usually, in profusion. This very interesting change is accompanied by a change in the outline of the leaves. They cease to be three or five angled, and all the new ones become more or less oval and acutely pointed. At the first blush it seems most surely a different kind of plant, and to deem it, and designate it when in this ripe and prolific condition, the "female ivy," was in the infancy of botanical science most natural, though erroneous.

In the *Comedy of Errors*, ii., 2, Adriana calls it "usurping ivy:"—

> If aught possess thee from me, it is dross,
> Usurping ivy, brier, or idle moss,
> Who all, for want of pruning, with intrusion
> Infect thy sap, and live on thy confusion.

Here the citation, excellent in its way, is metaphorical; the ivy is an interloper, and Adriana would have us believe it hurtful and a spoiler besides. She continues,—

> The time was once when thou, unurged, wouldst vow
> That never words were music to thine ear,
> That never object pleasing in thine eye,
> That never touch well-welcome to thy hand,
> That never meat sweet-savour'd in thy taste
> Unless I spake, look'd, touch'd, or carved to thee.
> How comes it now, my husband, oh, how comes it
> That thou art then estrangèd from thyself?

Ivy is again introduced in the *Tempest*, i., 2, now as a figure for ingratitude. Prospero is describing the wicked conduct of his brother Sebastian:—

> He was
> The ivy which had hid my princely trunk,
> And sucked my verdure out on't.

In both these latter passages Shakspere adopted the idea of his time (quite as natural of its kind, as that of the berried plant being "female"), that ivy both feeds upon the tree up which it clambers, and operates injuriously upon the sap. Neither idea has any real foundation. Ivy is not a parasite. It is not even an epiphyte. The root is in the ground, and like all other leafy climbing plants of temperate countries, it obtains its nourishment from the soil and the enveloping atmosphere. When injury is done to a tree (and this happens only to weakly

and exceptional individuals), the hurt comes of the tree
being too much shut in from the beneficent action of the
air, the rain, and the sunshine. Old trees observed to
be dead or dying in the midst of masses of ivy, were
in all likelihood beginning to die while the ivy about
them was still quite young, and would have died independently of it.   The ascent of the ivy up such trees
would be the same as up an old wall.

Ivy, in ancient times, was one of the plants dedicated
to Bacchus — "hedera est gratissima Baccho."   The
ground of the consecration was that when the child lay
in his cradle, an object of vengeance with Juno, the
nymphs of Nisa concealed him with ivy trails.   Homer
bestows on him the epithet of "ivy-tressed." Not that ivy
was purely the god's own.   When Calliope, lyre in hand,
sang that charming ode in praise of Ceres, she had her
hair bound with a sprig of the same beautiful evergreen.
The dedication to Bacchus is mentioned also in the
Second Book of Maccabees, where, when the temple is
desecrated, the Jews are compelled to go in procession,
carrying κισσός.   Hence, in the middle ages, it became
customary with the owners of inns and taverns to hang a
small bough of ivy over the entrance, by way of invitation to the thirsty traveller.   References to this custom
appear frequently in old authors, though not earlier,
perhaps, than in Chaucer.   A very curious and distinct
allusion to it occurs in the preface to the Latin translation of the famous old German History of Plants by
Hieronimus Tragus, published in 1532.   The Latin

version, by David Kyber, appeared in 1552. It is usual, says Kyber, "to descant upon the merits of an author one may edit or translate. But in the present instance we shall refrain from doing so, partly because this is already done by another hand (Conrad Gesner's) in the Introduction; partly because, according to the common proverb, Good wine requires no suspended ivy." Very naturally, the descriptive portion of the original proverb, whatever it was, would abbreviate itself into the smallest possible compass, becoming at last "Good wine needs no bush," as cited by Rosalind in the charming Epilogue to *As You Like It:*—

ROSALIND: It is not the fashion to see the lady the epilogue; but it is no more unhandsome than to see the lord the prologue. If it be true that good wine needs no bush, 'tis true that a good play needs no epilogue; yet to good wine they do use good bushes, and good plays prove the better by the help of good epilogues. What a case am I in then, that am neither a good epilogue nor cannot insinuate with you in the behalf of a good play! I am not furnished like a beggar, therefore to beg will not become me: my way is to conjure you; and I'll begin with the women. I charge you, O women, for the love you bear to men, to like as much of this play as pleases you: and I charge you, O men, for the love you bear to women—as I perceive by your simpering, none of you hates them—that between you and the women the play may please. If I were a woman I would kiss as many of you as had beards that pleased me, complexions that liked me, and breaths that I defied not: and, I am sure, as many as have good beards or good faces or sweet breaths will, for my kind offer, when I make curtsy, bid me farewell.

The name of "The Bush" has not even yet disappeared from English taverns.

The allusion in the *Winter's Tale*, iii., 3, "They have scared away two of my best sheep; if anywhere I find them 'tis by the sea-side, browsing on ivy,"—is not really Shakspere's, and applies to some quite different plant. The phrase is palpably taken from Greene's *Pandosto*, the novel upon which the play was partly founded, and in which we have — "It fortuned a poore mercenary sheepheard . . . missed one of his sheepe, and . . . wandered downe toward the sea cliffes to see if perchaunce the sheepe was browsing on the sea ivy, whereon they greatly doe feede," etc. Something very palatable is here intended. Ivy is not so. That ivy occurs near the sea is quite true, but under no circumstances can it be considered a maritime plant, and Greene must have been thinking of one to which sheep incline. The original *Pandosto* was published in 1588; the *Winter's Tale*, Shakspere's last complete play, was written in 1611, five years only before the poet's death. Seeing whence the allusion to sea ivy was derived, it may thus at once be relegated to the list, not very short, of Shaksperean anachronisms—no play showing less regard to accuracy in respect of time and geography, than the *Winter's Tale* itself. We are in no wise required to demur to these anachronisms; much less are we called upon to quarrel with them. Unpardonable in actual history, anachronisms, in the ideal, count with the very slightest of blemishes. The ideal is always young and is tied to no locality. The poets rightfully so called may always be known by their emerald wings, such as render them independent of epoch and country. In the *Winter's Tale* we seem seated upon the soft turf of some beautiful seclusion. Here the fascinating grace of Shakspere's youth "shakes hands with the thoughtfulness of his mature age." "The golden glow of his genius," says Mr. Furnivall, "is over it; the sweet country air all through it . . . As long as men can think, shall Perdita brighten and sweeten, Hermione ennoble, men's minds and lives . . . Its purpose, its lesson, is to teach forgiveness of wrongs, not vengeance for them; to give the sinner time to repent and amend, not to cut him off in his sin."

The Mistletoe.—Mistletoe, unlike ivy, is a genuine parasite, fixing itself upon the branches of trees of very various kinds, twenty different ones at least, in England alone,—though upon the oak, with which tradition so closely connects the plant, very seldom. The known examples at the present day can be counted upon the fingers. The favourite haunts of this curious plant are apple-orchards. Shakspere (if the lines be really his) connects it with the wilderness, where mistletoe occurs frequently, seated upon hawthorns. Tamora, Queen of the Goths, says in *Titus Andronicus*, ii., 3:—

> Have I not reason, think you, to look pale?
> These two have 'ticed me hither to this place,
> A barren, detested vale you see it is;
> The trees, though summer, yet forlorn and lean,
> O'ercome with moss, and baleful mistletoe.

Moss.—Moss is mentioned in three other places as an inhabitant of the bark of decrepit trees:—

> Under an oak whose boughs were moss'd with age.
>> *As You Like It*, iv., 3.

It appears also in *1st Henry the Fourth*, iii., 1, as an occupant of the walls of ancient and ruinous buildings,—

> Steeples and moss-grown towers.

With Shakspere the name denoted, as with all other poets, not merely the little green-leaved plants called *Musci* by the botanists, but the commoner and conspicuous kinds of lichen. These last quite as frequently drape the branches of aged foresters with their curiously tufted grey filaments; and still oftener in their horizontal, crispy, and lace-edged kinds, confer those beautiful time-stains, both grey and yellow, upon broken arch and mouldering turret, which catch the eye of the tasteful artist, and test the ingenuity of his pencil. To the minute observer they are unfailing objects of delicate pleasure. Their crowds of pretty cups "make glad the solitary place," and in winter particularly, show that nature has festivals for all seasons. The green-leaved or true mosses chiefly observed by Shakspere, would be species of the large and varied genus Hypnum, especially the *rutabulum* and the shining *sericeum*, though in one place he alludes more particularly to the elegant little furry cushions formed by others of this most pleasing race of nature's minims.

> With fairest flowers,
> While summer lasts, and I live here, Fidèle,
> I'll sweeten thy sad grave. Thou shalt not lack
> The flower that's like thy face, pale primrose; nor

The azured harebell, like thy veins; no, nor
The leaf of eglantine, whom not to slander
Outsweeten'd not thy breath.　The ruddock would
With charitable bill (O bill sore shaming
Those rich-left heirs that let their fathers lie
Without a monument) ! bring thee all this,
Yea, and furred moss besides.—*Cymbeline,* iv., 2.

Imogen deserves all.　A more loveable character is found
nowhere in Shakspere.　She discloses herself, once for
all, in assuming while persecuted, the name of "Fidèle."
How beautiful, too, the introduction of the little bird !
Men failing in their duty, the ruddock at least would not
forget, perhaps even be the first, to scatter flowers.
Not that the pretty superstition is of date no remoter
than the Shaksperean age, for although fastened to the
robin, perhaps not so very long before, kindly bird-work
of this general nature was believed in quite fifteen centuries antecedently.　Horace tells us that straying away
from home when a boy, and lying asleep in the wilds, he
was covered with leaves by the wood-pigeons *(Carm.*
iii., 4).　With Shakspere, nevertheless, it becomes new.
There are many things in objective nature, as well as in
the realm of poetry, which are at once curiously old and
delightfully young.　Look, for instance, at those crowds of
little sylvan equiseta, with their many cupolas of green,
plant-forms not so much of to-day as of the times when
they existed in company with sigillarias and lepidodendra,
and the thousand other quaint antediluvians one sees in
the museums.

The concluding line of this *Cymbeline* passage is

printed in the first folio, 1623, as above given, with a full stop at "besides,"—

> Yea, and furr'd moss besides.

Then resuming,

> When flowers are none,
> To winter-ground thy corse—

But in modern reprints the punctuation has been altered to—

> Yea, and furred moss besides, when flowers are none,
> To winter-ground thy corse.

It seems to have been thought that Shakspere meant that the same birds would go on in winter. But surely, as pointed out by Mr. Hunter, it is Arviragus. Arviragus begins with telling what he will do "while summer lasts." Then he is about to say what he will do in the time of frost and snow. He intends to imitate the gardeners when they "winter-ground" their particularly tender plants with litter,—

> When flowers are none,
> To winter-ground thy corse—(I'll)—

Guiderius, however, breaks in; will not let him proceed; stops him short with—

> Pr'ythee have done,
> And do not play in wench-like words with that
> Which is so serious.

So much for the sense and the proper stops. Expecting more sweet poetry, one is inclined, for the moment, to be vexed by the interruption, especially as Guiderius cares to talk only about the funeral. Let be. Shakspere knew what he was doing. We may thank him for reminding us, as in a thousand other places, that the best and truest sentiment is always that which flows forward into the faithful performance of the prosaic and practical duties of life; and that sentiment which fails to express itself in sound personal conduct, is like religion without charity, a tinkling cymbal. Sentiment and the practical are not, as many people think, antagonistic. The happiest conjugal unions ever witnessed in the world are those in which common-sense has for its partner the cheerful abundance of a serene imagination.

"Who can read," Mr. Hunter remarks, the words concluding his own volume,—"Who can read this beautiful scene, though it may be for the hundredth time, without taking from it the most sensible pleasure, and feeling most deeply grateful to the men of other days who, in the treasures of song, have sent down to us so much that administers such high and innocent delight? They have been the great and generous benefactors of mankind, scattering their gifts with an unsparing and an equal hand, strewing them in the humblest cottage, and laying them before princes in their gorgeous palaces; and worthy are they in turn to receive all honour, for the utmost that can be done for them is no equivalent for the extensive and enduring benefits which they have conferred upon their kind. I have in this work scattered a few flowers upon the grave of one of them."*

---

* New Illustrations, ii., 301.

Guiderius' desire that sentiment shall not excuse neglect
of the practical, is another of the Shaksperean ways of
illustrating what the master never forgets — the great
law of nature that trouble shall always be in some way
compensated; that no year shall be altogether dreary:—

> The time will bring on summer,
> When briers shall have leaves as well as thorns,
> And be as sweet as sharp;

that contraries in human experience shall never be far
asunder—a very ancient story, no doubt, but always open
to new and impressive illustration, since he—the master—
never forgets either that behind the sweetest and most
joyous things of earth there is always an undertone of
sadness, waiting the opportunity to express itself, and
which it is quite as much his bounden duty to recognise
and deal with. This is why, in almost all the greatest
of Shakspere's works, as in Homer, we have interfusion,
so skilful, of the bright and cheery with the grave and
solemn, of fun and playfulness with the philosophic, and
in most of the tragedies, amid agony and blood, or the
thirst for it, even the comic and the humorous. How
admirably, in the *Merchant of Venice*, are the stern and
terrible features of the scene in the judgment hall
relieved by those of the garden at Belmont:—

> SHYLOCK: So do I answer *you!*
> The pound of flesh, which I demand of him,
> Is dearly bought, is mine, and I will have it.
> If you deny me, fie upon your law !
> There is no force in the decrees of Venice.
> I stand for judgment:—answer, shall I have it ?

GRATIANO: O be thou cursed, inexorable dog!

. . . .

> The moon shines bright. In such a night as this,
> When the sweet wind did gently kiss the trees,
> And they did make no noise :—in such a night,
> Troilus, methinks, mounted the Trojan walls,
> And sighed his soul toward the Grecian tents,
> Where Cressid lay that night.

In *Hamlet*, after the same manner, we turn from crime, and madness, and the intensest pathos, to the mirthful chat of the old grave-diggers in the churchyard. In *Macbeth*, before the horrors begin, the mind is pleased, and in a measure invigorated, as if from some fragrant and sustaining wine-cup, by what is perhaps the loveliest picture of natural repose in the whole of our immortal poet:—

> DUNCAN: This castle hath a pleasant seat; the air
> Nimbly and sweetly recommends itself
> Unto our gentle senses.
> BANQUO: This guest of summer,
> The temple-haunting martlet, doth approve
> By his lov'd mansionry, that the heavens' breath
> Smells wooingly here; no jutty, frieze, or buttress,
> Nor coigne of vantage, but this bird hath made
> His pendent bed, and procreant cradle :—where they
> Most breed and haunt, I have observed, the air
> Is delicate.—(i., 6.)

Fairy-rings.—Upon out-of-the-way green rural surfaces—sunward slopes where the turf is short, and inlaid with eyebright and other *bijouterie*—there are very commonly seen broad dark circles, varying in diameter from a few feet to many yards. These are the "fairy-rings" alike of poetry and science, which when talking of them, does not

disdain the language of fable.   They are referable, as
before stated, to the centrifugal growth of certain fungi
of the agaricus or mushroom type, though, according
to Prof. Buckman, by no means always dependent upon
fungi for the initiative.   In the Elizabethan age, when
"fungi" and "mushrooms" meant the same thing, these
pretty circles were believed to indicate, when discovered
at sunrise, the spots where, the night before, the fairies
had been dancing; thus to have been formed, at once,
of their full dimensions.   Shakspere has several beautiful
references to them.   First, in the invocation of the spirits
by Prospero, one of the most exquisite minglings of
veritable nature and the unreal to be found anywhere
in his works:—

> Ye elves of hills, brooks, standing lakes, and groves,
> And ye that on the sands with printless foot
> Do chase the ebbing Neptune, and do fly him
> When he comes back:—you demi-puppets, that
> By moonshine do the greensome ringlets make
> Whereof the ewe not bites, and you whose pastime
> Is to make midnight mushrooms.

"Whereof the ewe not bites" is in acceptance of another
popular fancy of the time, sheep having in reality no
objection to eat the grass marked by the rings in
question.   "Greensome," again, almost always printed
"green sour," is surely the epithet intended.   "Green
sour" is by no means Shaksperean.   So doubtful is the
authenticity, that some have proposed to read "greensward."   "Greensome" has parallels in "winsome,"
"gladsome," "lightsome," and fifty besides.

Then comes the charming couplet in the *Midsummer Night's Dream*, ii., 1 :—

> And I serve the Fairy Queen,
> To dew her orbs upon the green,—

to moisten them, that is, with dew, after they have been worn dry by the tripping of the little dancers;—then in the *Merry Wives* v., 5, the queen's own personal injunction to her maids and the company, one and all, to be sure they take full vocal part in the entertainment :—

> And, nightly, meadow-fairies, look you sing
> Like to the Garter's compass, in a ring;
> The expressure that it bears, green let it be,
> More fertile fresh than all the field to see.

The "Garter" here referred to is the famous badge identified with the legend of *Honi soit qui mal y pense.* "Field" is also to be understood as involving an heraldic allusion, the "field" being, in armory, the surface upon which the "charges" are depicted.

Reeds and Rushes.—Reeds, the emblem, from time immemorial, of weakness and imbecility, in Shakspere take their old accustomed place, mention being made of them in *Antony and Cleopatra*, ii., 7; *Cymbeline*, iv., 2; and very elegantly in *1st Henry the Fourth*, i., 3, where we have the hand-to-hand combat of Mortimer and "great Glendower" :—

> Three times they breathed, and three times did they drink
> Upon agreement, of swift Severn's flood,
> Who then, affrighted with their bloody looks,

S

Ran fearfully among the trembling reeds,
And hid his crisp head in the hollow bank.

This beautiful river has a moment before been called
"gentle." Now, by a rich personification, after the
manner of the Hebrews, when they spoke of the laughing and singing of the cornfields, the reeds upon the
banks partake of the dismay of the stream and shiver
anew. The description is one of those so frequent in
Shakspere which we do not so much read as mentally
transfigure into life and visible action. In the highest
forms of art, the master is always known by his awaken-
ing a higher consciousness than the one immediately
addressed.

"Reed" being a generic or collective term, no particular species can be insisted upon either in this or
any other passage. But we shall not err in thinking of
Shakspere's reed as that tall and elegant plant, the
Arundo Phragmites, largest and most tropical looking
of the English grasses, and familiar in its impurpled and
downy plumes, the ornament of the marshes throughout
the fall of the year, the more particularly since the reference in the *Tempest*, v., i., to the use of the reed for
thatch* points to this plant almost definitely.

Portia's words, *Merchant of Venice*, iii., 4,—

And speak, between the change of man and boy,
With a reed voice,—

refer to the shrillness of tone characteristic of that period

---

* After the manner of old Baucis and Philemon's cottage—
stipulis et canna tectâ palustri.—Met., viii., 630.

of life, image of the whistled melody of the most ancient
of musical instruments, shaped, quite possibly, from the
Phragmites; in any case, from the still more serviceable
as well as more stately *Donax.*

> Sweet, sweet, sweet, O Pan !
> Piercing sweet by the river.

Apart from the beauty of the myth of the origin of the
first musical pipe, the incident of the employment of
the reed for its manufacture is one of the most interesting in literal history, since in this we find the first step
towards the glorious Organ of to-day. Complex as it is
now, the organ began with the discovery that air may
be forced into a closed cavity, such as that of the reed,
and then, with ingenuity, and by means of lateral open-
ings, be forced out. Poetry tells the tale in its own way,
but virtually this is just what happened one summer's
eve in old Arcadia.

Rushes are cited in various association. In two or
three out of the total of sixteen passages in which the
word occurs, they appear after the same manner as
reeds, or as emblems of instability and trifles:—

> Our gates,
> Which yet seem shut, we have but pinned with rushes;
> They'll open of themselves.—*Coriolanus*, i., 4.

In one instance *(All's Well*, ii., 2) there is an allusion
to the scandalous mediæval custom of marrying, or rather
pretending to marry, with a ring made of a rush. In
another, they supply wicks for candles *(Taming of the
Shrew*, iv., 5); the remainder bear chiefly upon the

employment of rushes, before the introduction of carpets, for strewing upon the floors of the houses of people of rank, a practice very frequently mentioned by old authors, as in the piteous description of the unhappy queen of Edward IV. *(née* Elizabeth Woodville), when news was brought her of the monarch's death:— "She sat alow (alone) on the rushes, all desolate and dismayed." Grumio, when he asks, "Is supper ready?" adds, "are the rushes strewed?" *(Ibid,* iv., 1). In *1st Henry the Fourth,* iii., 1, the lady's Welsh, made by her tongue and "swelling heavens,"

> Sweet as ditties highly penned,

is translated for Mortimer,

> She bids you,
> Upon the wanton rushes lay you down,
> And rest your gentle head upon her lap,
> And she will sing the song that pleaseth you;
> And on your eyelids crown the god of sleep,
> Charming your blood with pleasing heaviness.

In the *Rape of Lucrece,* and in *Cymbeline,* ii., 2, Shakspere adverts to the vile incident in the life of Tarquin which deservedly lost him his kingdom,—

> Our Tarquin thus
> Did softly press the rushes ere he wakened
> The chastity he wounded.

Here he places before us, in the most vivid manner, the simplicity of the times when old Rome had kings, the luxury of the empire not arriving for many ages afterwards. There is no evidence that England was anticipated in this matter by ancient Rome. The practice, nevertheless, was of indefinitely early date. In Theocritus (xiii.), when the Argonauts have gone on shore in the land of the Cyanians, and the nymphs, enamoured, steal Hylas from his companions, the sailors are described as strewing the floor with the leaves of the butomus—not the plant so called to-day, but some large carex or cyperaceous plant. That in the fourteenth century the custom was continental, is proved by the allusions in Froissart. The early Italian authors refer to it in their word *giuncare.*

The rushes thus employed are intended also by Shakspere, by implication, when the presence-chamber, or briefly, the "presence" is said to be strewed. In those admirable lines, for example, addressed to an individual, but, like everything else noble and good in Shakspere, intended for everybody, when John o' Gaunt, "time-honoured Lancaster," reminds us that sunshine is at the command of every one, and that happiness is not a wonderful diamond, to be sought afar off, but the multiplying and creative power of a thankful heart:—

> All places that the eye of heaven visits,
> Are, to a wise man, ports and happy havens.
>
> .   .   .   .
>
> Look, what thy soul holds dear. Imagine it
> To lie the way thou go'st, not whence thou cam'st—
> Suppose the singing birds musicians ;
> The grass whereon thou tread'st, the presence strewed ;
> The flowers, fair ladies ; and thy steps no more
> Than a delightful measure, or a dance ;

For gnarling sorrow hath less power to bite
The man that mocks at it, and sets it light.
                              King Richard the Second, i., 3.

The kinds of rush employed would be ordinarily the
common *junci* of rough waste land. Bulleyn says somewhat definitely, "Rushes that grow upon drie grounds be
good to strewe in Halles, Chambers, and Galleries to
walke upon, defendyng apparell (as traines of gownes
and kertles) from duste." "Rushes," he quaintly adds,
"be olde Courtiers; and when they be nothing worthe,
then they be caste out of the doores. So be many that
dooe tread upon them." But though "drie ground"
rushes may have had the preference, we cannot suppose
neglect of the abounding denizens of the watery hollow,
the *Juncus effusus* and its allies, always so friendly with
the pink flos-cuculi and the ivory meadow-sweet. Use
would probably be made also of the spongy stalks of the
bulrush, *Scirpus lacustris,* and perhaps to some extent,
of the cinnamon-rush or sweet-flag, *Acorus Calamus.*
"The sweet-scented rush," says Mr. Ellacombe, was
always used when it could be procured. Loudon, in
1829, and Dr. Lindley, in 1833, make the same state-
ment, to which the Rev. C. A. Johns adds, in his
Flowers of the Field, that "as it did not grow near
London, but had to be fetched, at great expense, from
Norfolk and Suffolk, one of the charges of extravagance
against Cardinal Wolsey was that he caused his floors to
be strewed with rushes too frequently." The fresher
they were, so much of course the cleaner and more

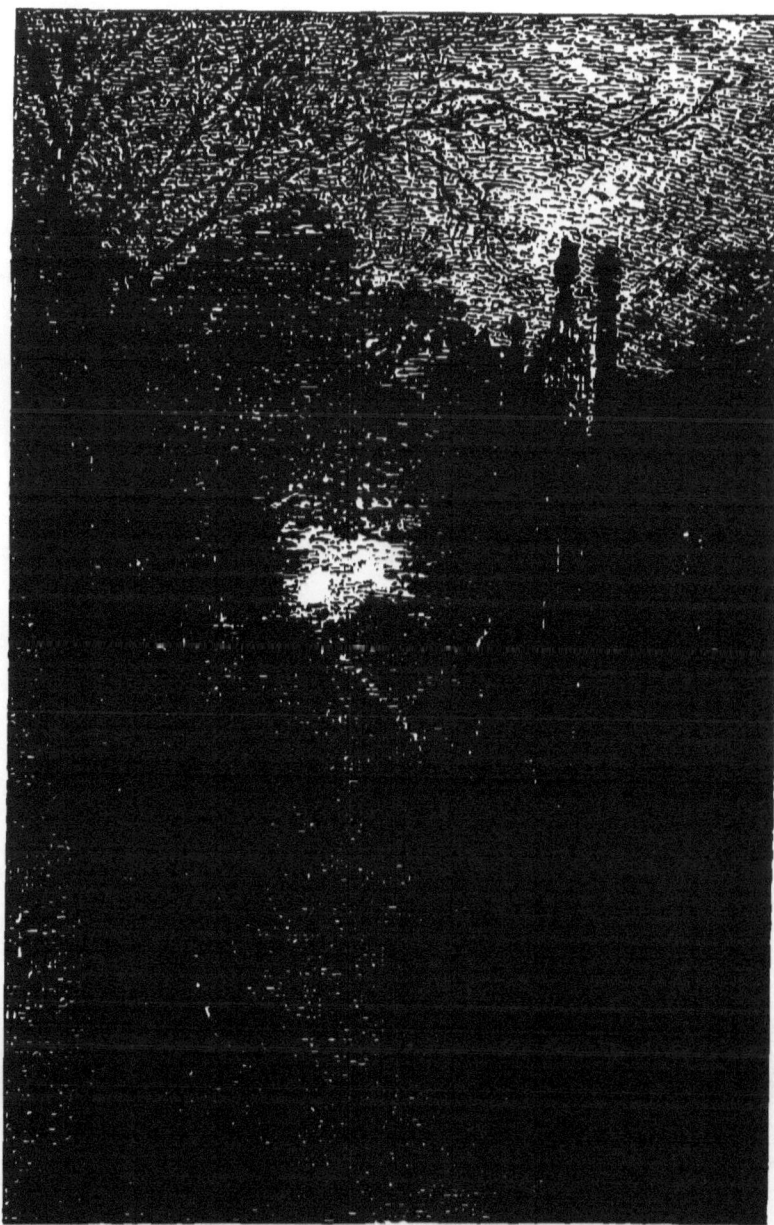

Drawn & Etched by Tho Lotherbrow

pleasant, especially for guests, whom it was customary to honour with new ones:—

> Where is this stranger? Rushes, ladies, rushes,
> Rushes as green as summer for this stranger.

Poor Wolsey, the end not less majestic than the reign, though faultful, never forget how much England owes to him. When, temp. Henry VIII., endeavour was made to introduce the language of ancient Greece into the Universities, it was met by the most furious opposition. Erasmus' Greek New Testament (the first of its kind) was denounced and reviled. Wolsey came to the rescue, and Wolsey we may thank for giving to Greek the splendid anchorage in England which was never afterwards interfered with. Erasmus himself bears eloquent testimony to Wolsey's merits in the matter. "Learning," he says, "as yet struggling with the patrons of the ancient ignorance, he upheld by his favour, defended by his authority, adorned by his splendour, and cherished by his kindness."

The authorities as to the use of the acorus stand high. There is a little misgiving over it, nevertheless, since the plant is not mentioned in the old herbals as a wild English one, and would seem not to have become common in England until after the Restoration. It is to be observed also, that neither Shakspere, who was not the man to miss anything characteristic, nor any of his contemporaries, ever speak of their rushes as odoriferous, not even when used like Romeo's, for the floors of dancing saloons:—

Let wantons, light of heart,
Tickle the senseless rushes with their heels.—(i., iv.)

Botanically regarded, a plant more interesting does not
exist.   Growing just within the margin of ponds, to reach
it without wetting the feet is not easy—faint heart never
won fair lady.   The leaves, which come up in crowds,
are tall, vertical, flat, narrow, and curiously crimped
along one edge.   The flowers, borne upon stems of
similar figure, constitute curious lateral tapering cones,
not very unlike the spires of churches; the colour a
pretty greenish-brown; the surface covered with minute
golden mosaic.

Shakspere's "Sedges."   Sedge, a term little altered
from the Anglo-Saxon, must be understood as denoting
in poetry any hard and wiry vegetation that may border
the pool or the river-side.   These are the natural haunts
of the larger English species of botanical "sedge," a
dozen perhaps in all, and which are themselves haunted,
in turn, by various birds, the sedge-warbler to wit.
Hence, in *Much Ado*, ii., 1,—

Alas, poor hurt fowl; now will he creep into sedges.

Four other allusions show how easily the word can help
to make beautiful pictures:—

The gentle Severn's sedgy bank.
1st Henry the Fourth, i., 3.

You nymphs, called naiads, of the winding brooks,
With your sedged crowns, and ever-harmless looks.
Tempest, iv., 1.

And Cytherea, all in sedges hid,
Which seem to move and wanton with her breath.
                    Taming of the Shrew.—Introd.

The current that with gentle murmur glides,
Thou know'st, being stopped, impatiently doth rage.
But when his fair course is not hinderèd,
He makes sweet music with the enamelled stones,
Giving a gentle kiss to every sedge
He overtaketh in his pilgrimage ;
And so by many winding nooks he strays
With willing sport, to the wild ocean.
                    Two Gentlemen of Verona, ii., 7.

" Pilgrimage " is perfect, the idea being that the trickle on the mountain-side sees its Jerusalem or Mecca from afar, never thenceforward wavers, is faithful to the end.

There are a few Shaksperean occupants yet of the wilderness and the wayside. In the *Merry Wives,* v., 5, mention is made of " bilberries," the violet-black fruit of the *Vaccinium Myrtillus,* so common upon the moor-lands, and more usually called whortleberries. Staining the lips with a deep purplish-blue when eaten in quantity, they supply an apt colour comparison. Whortle was very naturally contracted into hurtle. Hence, in heraldry, the deep blue roundlets occasionally met with received the name of *hurtes,* just as the green roundlets were designated pommes.

In the *Tempest,* ii., 2, where odious Caliban becomes for a moment endurable, the earth-chestnut is adverted to :—

I, with my long nails, will dig thee pig-nuts.

Shakspere here again gives preference to the less usual,

though quite as suitable English name of that very pretty and interesting plant, the *Bunium flexuosum* of the botanists. In England, in its way, the pignut is unique. The aërial portion reminds one of hemlock, of which, indeed, this plant is a near ally, only that every part is in miniature, and that the umbels, before blooming, are pendulous. The usual height of the stem is about fifteen inches. Below the surface of the ground it bends to and fro in the most curiously irregular manner, diminishing in thickness at every turn, till at last we reach the round brown nut, white inside, hard, and of pleasant nutty flavour. There is no pulling it up by force. Shakspere had learned in his boyhood that the pignut is won alone by patience and perseverance. Writing the line, he was a boy over again. Say rather that the man of genius in spirit never ceases to be a boy.

Another schoolboys' plant is plainly intended in *1st Henry the Fourth*, ii., 4, "Yea, and to tickle our noses with speargrass to make them bleed." Whether the allusion be to the green needles of the horsetail, or to the feathered leaves of the common milfoil, *Achillea Millefolium*, by Lyte called "nose-bleed," is a point yet to be settled.

In i., 3 of the same play there seems to be an allusion to the "shepherd's purse," *Capsella Bursa-pastoris:*—

> My liege, I did deny no prisoners.
> But, I remember, when the fight was done,
> When I was dry with rage, and extreme toil,
> Breathless and faint, leaning upon my sword,
> Came there a certain lord, neat, trimly dressed,

Fresh as a bridegroom.

.   .   .

He made me mad
To see him shine so brisk, and smell so sweet,
And talk so like a waiting gentlewoman,
Of guns, and drums, and wounds (God save the mark)!
And telling me the sovereign'st thing on earth
Was parmaceti, for an inward bruise.

The immediate reference here is to spermaceti, in the Shaksperean times a famous remedy for hurts. But the capsella had a similar reputation—"the juice," says the Grete Herbal, "when drunk, staieth bleeding in any part of the body." Hence it got the name of "Poor man's Parmaceti," with involved equivocal allusion to the poor man's queen of medicines, the well replenished purse.

The common duckmeat or duckweed, *Lemna minor*, the little plant of the thousand green spangles, which roofs all rustic stagnant waters, is introduced in *King Lear*, iii., 4, "the green mantle of the stagnant pool." The "vagabond flag upon the stream," of *Antony and Cleopatra* i., 4, means, probably, the surface plants in general of slow waters, or such as with changing conditions, glide to and fro. The leaf of the common plantain, *Plantago major*, anciently reputed good for a broken shin, is referred to in *Love's Labour's Lost*, iii., 1, and in *Romeo and Juliet*, i., 2.

When, however, in *Troilus and Cressida*, iii., 2, we have—

As true as steel, as plantage to the moon—

no reference is intended to the plantago in particular. The word is here a kind of synonym of "herbage," denoting vegetation in general, the mysterious lunar influence upon so large a portion of which, with a sort of reciprocal fealty to the moon on the part of the plant, constantly crops up in the superstitions of the middle ages.

Chapter Thirteenth.

THE MARKET-PLACE AND THE SHOPS.

> We, ignorant of ourselves,
> Beg often our own harms, which the wise powers
> Deny us for our good ; so find we profit,
> By losing of our prayers.—*Antony and Cleopatra*, ii., 1.

S with all the cultivated fruits and the commonly cultivated esculent vegetables mentioned by Shakspere, so with a considerable number of plant products sold in the market-place and the shops, some for food, others for medicinal use, others again to subserve various economic purposes. The allusions very seldom carry with them anything either poetical or inviting, except in so far as they are humorous or facetious; and here let us not forget that to Shakspere the most distasteful of all men, not criminal, was that

one who after incapacity for delight in music,* could not laugh, and laugh heartily. That these humorous and facetious allusions should occur is thus quite in keeping with the genial and stimulating humanity of Shakspere. Without them, the plays would not have shown him truly, and we should have lacked a great portion of the delightful influence which, whether it makes us laugh, or weep, or meditate, is still, in its way, inestimable. Shakspere's best characters are in every instance joyous themselves, and most of them are able to be merry; and perhaps it is in the degree that we strive to assimilate ourselves to those matchless creations that we rise in the scale of existence—not forgetting that the central feature of all those best characters is identical with that of the man who fears God and obeys His law. The nature of these references renders it unnecessary to dwell upon them at length;—they so often speak for themselves that, in many cases, it is enough to quote the passage, or to indicate its position in the drama. Very interesting light

---

* CÆSAR: Let me have men about me that are fat;
Sleek-headed men, and such as sleep o' nights.
Yon Cassius has a lean and hungry look.
He thinks too much; such men are dangerous.
                    He loves no plays,
As thou dost, Antony—*he hears no music.*
Seldom he smiles, and smiles in such a sort
As if he mock'd himself, and scorned his spirit,
That could be moved to smile at anything.
Such men as he be never at heart's ease,
Whiles they behold a greater than themselves,
And therefore are they very dangerous.—*Julius Cæsar,* i., 2.

is thrown by the various citations upon the market business of the time, and still more upon the early history of English commerce, so far as regards articles imported. The trade carried on by the original English merchants would be chiefly with the maritime cities of the Mediterranean, those, in particular, of Spain, Italy, and the Levant. We need think only of Antonio and Bassanio, of unrelenting Shylock, and of graceful, wise, and witty Portia, to remember that the great centre of European mercantile fame for many ages, was Venice. Up to the time of Henry VII., the productions of India and other oriental countries would come by the same routes, in the first instance, as to Palestine in the days of Solomon and the Queen of Sheba. The discoveries made by Columbus, and the finding of a sea-passage to the East Indies by way of the Cape, events almost contemporaneous, gave quite a new character to foreign trade. The pride of Venice, as regarded oriental productions, passed in a great measure to Lisbon. The Portuguese merchants, in turn, made an intermediate emporium of Antwerp. The mercantile importance of the last-named city, in the time of Elizabeth, is described in the liveliest manner in the well-known pages of Ludovico Guicciardini ; and from Antwerp, in all likelihood, the Indian spices, and other such articles named by Shakspere, would now be proximately received, though the fruits ripened in the Mediterranean region would still arrive by way of the Bay of Biscay. Let us first enumerate these.

SHAKSPERE'S FOREIGN FRUITS.

As grapes take precedence in the garden, the corresponding place in the new series may be allowed to raisins, one of the most ancient of the historical fruits, and mentioned in the *Winter's Tale*, iv., 2, as raisins "of the sun," the commoner sorts being prepared by drying in ovens. The name is no more than an altered form of the Latin *racemus*, a bunch of grapes, as in Virgil:—

> Aspice ad antrum
> Silvestris raris sparsit labrusca racemis.
> (See how the wild vine with clusters here and there
> hath mantled over the grotto.—*Ecl.* v. 7.)

When Shakspere wrote, the word "reasons" was pronounced "raysons." Hence, the very natural *double entendre*, an instance of which occurs in the now proverbial line in *1st Henry the Fourth*, ii., 4, "Give you a reason on compulsion! If reasons = raisins were as plenty as blackberries, I would give no man a reason upon compulsion."

Currants.—The little Corinthian grapes imported from Zante, which by and by became "Corinths," then "corantes," and eventually "currants," are mentioned in the same passage in the *Winter's Tale*, iv., 2. The vine producing them is scarcely distinguishable from the plant which yields grapes *par excellence;* the flower-clusters, the bunches, and the berries alone are smaller. What a commentary upon the vicissitudes of nations, that Greece, once so illustrious, the parent of temple-architecture, of sculpture, oratory, history, poetry, and the drama, should

for ages before Shakspere talked of them, descend to the
exportation only of miniature raisins ! The red and white
currants of the garden, *Ribes rubrum*, though cultivated
in the time of Elizabeth, are not referred to by
Shakspere.

Dates, of celebrity almost as ancient, make their
appearance in the same passage (*Winter's Tale*, iv., 2),
from which, as also from the allusion in *Romeo and Juliet*,
iv., 4, it may be inferred that this fruit was a favourite
ingredient in confectionery. The same appears from
Parolles' uncivil comment in *All's Well*, i., 1, through the
characteristic and graceless medium of a verbal quibble,
upon grown-up ladies not yet married,—"Your date is
better in your pie and your porridge than in your cheek."
From Parolles such language comes quite consistently.
Mr. Halliwell describes him as "amusing enough; a fop,
nevertheless, a fool, a liar, a braggart, every way a knave,
too contemptible for anger." Any one of these attributes
is quite enough to cancel that noblest social capacity and
privilege of man—respect for woman, and courtesy at all
times, in addressing her. Shakspere, when speaking from
his own heart, is never rude to the unmarried woman.
He was not one of the thoughtless class who imagine
that an "old maid" can only be so by reason of some
personal defect, which rendered her unattractive in the
bygones. Many an old maid is so because bereaved in
the lang syne of her heart's idol, the loved and lost,
never spoken of, but never forgotten; to whose sacred
place she does not care to admit a successor. Many

T

another is what we find her because too high-minded, too *recherché* in her gentle and educated tastes, to have been satisfied with any one of her so-called "offers." She has never *condescended* to marry. For although it may be true, on the one side, that when a sensible and worthy man asks a woman to become his wife, he pays her the highest compliment in his power;—no man is entitled to suppose that condescension in the matter of marrying is a prerogative of his own gender. It is not to wives and mothers only that honour pertains. What more loveable sight in the world than that of the middle-aged unmarried daughter who keeps house for an aged and feeble father, who soothes his declining years, over-flowing with all dutiful affections, prays for him, and would lay down her life for him? The most repulsive female characters in Shakspere are the two ungrateful daughters, Goneril and Regan (in *Lear*);—the most heavenly one is their unmarried sister Cordelia.

Not that Shakspere was an admirer of feminine celibacy. Varied as are the stories of the plays, those, again to quote Mr. Halliwell, "in which the genial sunshine of his warm heart is invested, where his humanity becomes tender, and his imagination glowing, are his love-plots,"—those, in a word, which are designed to end in matrimony. When he wishes to describe, in a single phrase, the characteristics of woman after she becomes a wife, they are "her sweet perfections" (*Twelfth Night*, i., 1). No principle is more constantly illustrated by Shakspere than that the reciprocal and active love of men

and women originates, and assures, and gives completion
to what Goethe calls their Blessedness. Whether designed
or accidental, it is impossible to determine, but the fact
remains that one of the most profoundly original characters in the whole of Shakspere, as regards substantial
feminine love, is that one whose sweet name implies that
she Blesses—light-hearted, mirthful, sprightly Beatrice,—
Beatrice, who, as a maid, could not endure to be left out
of anything that was going on in the way of lively
business, such as was calculated to give pleasure to
those around her. Following close upon this comes the
illustration on the feminine side, of the transfiguring power
of love. Rosalind is a girl up to the time when she
becomes alarmed for Orlando's safety. Then her simple
interest in him changes on the instant into yearning
emotion, one remove only from the marital.

To put anything of our own into Shakspere, or to
make him appear to mean more than he intends, would
be taking a liberty, as already said, beyond excuse or
justification. It is legitimate, nevertheless, to consider
that whatever he did say, was said designedly. Granting
this, the circumstance is noteworthy that while nearly
every one of his principal female characters has a Christian
name that can scarcely be accounted English—necessarily
so, in most cases, because of the foreign scene of the
play—Portia, to wit, Miranda, Juliet, Ophelia, Hermione,
Helena, Hero, Viola, Perdita, Desdemona;—when he *does*
want an English name, and this for the type of a good,
modest, courteous, and loveable woman of the middle

social rank, neither princess nor plebeian, he goes for it
to his own wife. "Sweet Anne Page" is distinguished
also for her common-sense,—"Good mother, do not
marry me to that fool."

Almonds are mentioned in *Troilus and Cressida*,
v., 2:—"The parrot will not do more for an almond,"
words probably conveying some kind of proverbial
signification, now forgotten. The profound antiquity of
the esteem in which this fruit was held, beginning, as
was natural, in one of its native countries, the "land of
Israel," is illustrated in the history of Joseph and his
brethren. Allusions to it in classical literature are
scanty, nevertheless, though the singular beauty of the
tree, when in full bloom, meets with occasional notice;
and this may have been known to Shakspere, who was
not the man to overlook anything that would attract the
author of the *Faery Queene,*—

> Like to an almond-tree ymounted hye
> On top of green Selinis all alone,
> With blossoms brave bedeckèd daintily.—I., 7, xxxii.

In the Elizabethan age almonds were much in demand
for the making of "marchpanes," the antetype of the
modern macaroon, a sweet very evidently esteemed alike
by high and low, since in *Romeo and Juliet* we have from
one of the latter class, thoughts of his "Susan" notwithstanding,—"Save me a piece of marchpane" (i., 5).

The Orange, though not comparing with the almond in
respect of historical fame, had by the time of Shakspere
become quite as common in southern Europe, whence it

would seem to have been abundantly imported, for in *Coriolanus*, ii., 1, though the scene is laid in Rome, we are of course to understand a reference to London ways and characteristics, including the litigant street "orange-wife." No mention of this queenly fruit occurs in the classical authors, the "golden apples" of ancient mythology having been quinces; nor is there any notice of it in literature before the time of the Arabian authors of the tenth or eleventh century, with whom the orange first appears. The original conveyance of the tree from northern India may be assigned to a period some little earlier. Very soon after this it was carried to Sicily and Spain, and in due course to Italy and the south of France, the diffusion coming partly of Moorish conquest, partly of the old Venetian commerce, and in some degree, probably, of the enterprise of homeward-bound Crusaders, though in matters of this nature we must be careful not to accredit those steel-clad warriors over-much. The very name of the fruit is oriental, being no other, Max Müller tells us, than the Arabic *naranj*, which represents in turn an Aryan word. The initial *n* was retained in the original Latinised name, *anarantium*, afterwards changed, apparently because of the golden hue, into *aurantium*, and from this last, the transition, in France, into "orange," was easy and natural. The loss of the eastern initial *n* has a curious counterpart in the history of the name of the adder, which is properly, by derivation, "a nadder." When the fruit was first brought to England is not known. The earliest mention of it, as regards this country, appears to be in Queen Eleanor's household expense book for A.D. 1290, where it is said that she purchased from a Spanish ship which came to Portsmouth "vii poma de orenge."

In the *Midsummer Night's Dream* the hue of the rind helps upon two occasions to describe a colour, very happily, in particular, in Bottom's little song about the birds:—

> The ousel-cock, so black of hue,
>   With orange-tawny bill;
> The throstle, with his note so true,
>   The wren with little quill.—(iii., 1.)

"Quill," a word ultimately derived from *calamus*, a reed, refers to the sweet, high-pitched, piercing note, similar to the reed-music heard that immemorial summer's eve when

> The sun on the hill forgot to die,
> The lilies revived, and the dragon-fly
>   Came back to dream on the river,—

the note which Shakspere had plainly learned so well how to distinguish amid the songs of other birds, and which for our own parts always make one wonder how so much sound can be poured forth from so small a body. The other mention of this colour occurs in i., 2, "Your orange-tawny beard." It seems to have been considered one of ill-promise in mankind, since Beatrice, in *Much Ado*, playing upon Seville, the name of the Spanish city whence the fruit was partly received, says jocularly of Count Claudio, that he is "civil as an orange, and something of that jealous complexion" (ii., 1).

The Lemon is mentioned once, slightly, in *Love's Labour's Lost*, v., 2. Like the orange, it came originally from the forests of India, arriving in Europe, probably, in the same way and about the same time.

The Pomegranate, also mentioned once, is cited, we may be sure, from personal knowledge, acquired in all likelihood in the metropolis, though the scene is laid in Italy, occurring in *All's Well*, ii., 3. Contemplating the history, we are thrown back, yet again, into the remotest antiquity, Hebrew scripture and classical fable alike supplying curious illustrations of the primitive fame. Carried westwards at a very early period, the Romans seem to have received this fruit first from Carthage, queen Dido's own once glorious city: the Arabs, at a later period, conveyed it to the Spanish peninsula, and thence, or from Italy, it would seem to have reached England as early as the Anglo-Saxon times. The pomegranate was one of the fruits which, like the lily and the rose, travelled arm-in-arm with Christianity. The old French name, *pomme-granate*, "the grained apple," passed, after the conquest, into England, superseding the Anglo-Saxon *æpl-cyrnlu*. It was partly translated, however, into "apple-garnade," as in an old poem of about A.D. 1360, upon the destruction of Sodom and Gomorrah, in which the trees growing on the borders of the famous Sea are described as covered with produce as fair to look at

> As orange and other fryt and apple-garnade,
> Al so red and so ripe and rychely hwed.

The pomegranate-*tree* is alluded to in the immortal chamber-window scene—that one in *Romeo and Juliet* where we are shown so exquisitely how hard it is for a woman whose whole heart is enveloped by the beloved, to say "good-bye," and how the merest gossamer will suffice her for an argument why not to go:—

> JULIET: Wilt thou be gone? It is not yet near day:
>   It was the nightingale, and not the lark
>   That pierced the fearful hollow of thine ear:
>   Nightly she sings on yon pomegranate tree:
>   Believe me, love, it was the nightingale.
> ROMEO: It was the lark, the herald of the morn;
>   No nightingale. Look, love, what envious streaks
>   Do lace the severing clouds in yonder east.
>   Night's candles are burnt out, and jocund day
>   Stands tip-toe on the misty mountain-tops.—(iii., 5.)

In *Othello*, i., 3, we have the phrase "luscious as locusts." These are the dried fruit of the very celebrated tree of the Levantine region called the *Ceratonia Siliqua*, and commonly the "carob," whence also the name of "alga-roba beans," given to the pods, which are seven or eight inches long, narrow, curved, and flattened. The leaves are pinnate; the clustered pink flowers are incon-spicuous. From time immemorial these pods have been esteemed in their native countries, as a saccharine and nourishing food for brute creatures, though too coarse for mankind. They would appear to have become an article of very early import, though for what distinct purpose is not clear. At the present day they are used in the manufacture of molasses and of artificial cattle-food. Historically, carobs are interesting as having been the "husks" of the Parable of the Prodigal Son.

Of miscellaneous market-place edibles, not derived from the home kitchen-garden, about half-a-dozen kinds are mentioned by Shakspere, two of them, samphire and eryngoes, prepared from productions of the sea-shore. "Eryngoes," named in the *Merry Wives*, v., 5, were the candied roots, once greatly esteemed, of the *Eryngium maritimum*, grey denizen of the sea-side sandhills— their veritable "touch-me-not," distinguished at once by its intense prickliness, whence the synonym of "sea-holly," and the large egg-shaped heads of sky-blue flowers. The roots run to a great length among the sand; they are charged with a sweetish juice; and when softened in syrup, were in the Elizabethan age greatly valued, for purposes in harmony with Falstaff's character, as specified by Turner, p. 215 (1568).

Samphire, *Crithmum maritimum*, grows never upon the sandhills, but exclusively upon rocks and cliffs, such as are daily splashed by the salt water. The leaves are many-fingered and very succulent; the flowers are produced in yellowish umbels, the plant becoming a mass, in all, of some six inches in height. The juiciness and pleasant saline flavour recommended it, at a very early period, for use as a pickling vegetable. The collecting is at all times somewhat perilous. As a rule, samphire is often quite inaccessible except by means such as those referred to in the inimitable description in which it is mentioned,—suspending a man by a rope from the brow

of the cliff,—this particular cliff the most interesting in
our own island, since it is at Dover, and looks out upon
the channel:—

> How fearful
> And dizzy 'tis, to cast one's eye so low !
> The crows and choughs, that wing the midway air,
> Show scarce so gross as beetles.   Half-way down
> Hangs one that gathers samphire; dreadful trade !
> Methinks he seems no bigger than his head.
> The fishermen, that walk upon the beach,
> Appear like mice; and yon tall anchoring barque
> Diminished to her cock; her cock, a buoy
> Almost too small for sight.   The murmuring surge
> That on the unnumbered idle pebbles chafes,
> Cannot be heard so high.   I'll look no more,
> Lest my brain turn, and the deficient sight
> Topple down headlong.—*King Lear*, iv., 6.

Upon sea-side mud-flats there grows very commonly
another plant suitable for pickling, the salt-wort, *Sali-
cornia herbacea*, a singular production, made up of little
round, green, erect, and juicy pencils.   This, sometimes
called samphire by mistake, is by no means to be
thought of as Shakspere's samphire.

In the same scene in the *Merry Wives*, where Falstaff
invokes eryngoes, he prays also for a rain of " potatoes."
These were not the potatoes of to-day, which in the
Shaksperean age were only just beginning to be known
through introduction from South America by Sir Walter
Raleigh.   Falstaff's were the tubers of an East Indian
species of convolvulus, since named *Convolvulus Batatas*.
Large, solid, and very sweet, with great repute of pro-
perties similar to those of the eryngo, they became an

object of culture in southern Europe during the early middle ages, particularly in Spain, from which country those sold in the markets of the Elizabethan times would be obtained. They are now seldom heard of, and even the plant is scarcely known out of botanic gardens.

Most of the principal spices are named in the plays. Cloves appear in *Love's Labour's Lost*, v., 2. Nutmegs in the same scene, and in two other places, in one of which *(Winter's Tale*, iv., *2)* mace also is mentioned. Pepper appears in *Twelfth Night*, iii., 4, and again in two other places, only that now we must remember that with Shakspere the word was equivalent to " spice," as in *1st Henry the Fourth*, i., 3, " pepper-gingerbread." Ginger is named more frequently than any other substance of the kind. Observe that when in the *Winter's Tale*, iv., 2, a " race " of ginger is mentioned, it intends a *racine*, or portion of the root, as commonly used; whereas in *1st Henry the Fourth*, ii., 1, two " razes " of ginger mean two parcels or packages of it, " to be delivered at Charing Cross." To " knap " ginger, as in the *Merchant of Venice*, iii., 1, is to break it, as in the Biblical phrase, " He knappeth the spear in sunder."

Great regard seems to have been paid also in the Shaksperean age to Saffron, so noted for the delicate and very pure amber-yellow given out upon infusion in water, and which recommended it as a colouring ingredient in certain kinds of confectionery, and at one period for staining foppish neckties. Botanically, saffron consists of the dried stigmas of one of those beautiful species of crocus, with violet-purple flowers, which blossom in the autumn,—technically, the *Crocus sativus.* The distinguishing feature is that the stigmas, when the flowers are in perfection, hang out in the form of a deep golden-orange-coloured tassel between two of the petals. Indigenous to south-western Asia, saffron acquired celebrity at an exceedingly early period, and even more as a perfume than a condiment. Luxury employed it in various ways for the surprise and delight of guests at entertainments, and hence it falls quite naturally into the list of odoriferous things valued by Solomon, a monarch who seems to have been distinguished not more for his magnificence, his wisdom, and his enterprise, than for his predilection for odoriferous plants, as indicated in various passages in the Canticles; where to the sweets of the garden of spices, saffron, myrrh, and aloes, are added the fragrant walnut-tree and the snowy Lawsonia. That with the Greeks it should give name to the colour resulting from the infusion in water, was perfectly natural. It was present to the mind of old Homer when he spoke of "crocus-robed Aurora," the morning sky, in his poetic country, presenting this identical lovely tint,—

The cheerful Lady of the Light, dress'd in her saffron robe,
Dispers'd her beams in every part of this enflowerèd globe :—

with the Greeks it was no less natural that saffron should be deemed the fitting colour for the apparel of youthful and virgin terrestrial ladies, especially upon their wedding-days; girls wore saffron also upon occasions of profound solemnity, as when Iphigenia, about to be sacrificed, smote every one around with the "piteous glance of her eye."*

So vindictive can destiny be, that a saffron-coloured beard was traditionally ascribed to dissembling Judas! The arch-traitor was said to have gone shares in this matter with the first slayer of his kind. Hence, in the *Merry Wives*, i., 4, "No, forsooth; he hath but a little wee face, with a yellow beard,—a Cain-coloured beard."

The use of Saffron as a colouring ingredient in "warden-pies," is mentioned in the *Winter's Tale*, iv., 2. The delicious old Homeric picture is imitated in the *Tempest*, iv., 1, when Ceres invokes Iris,—

> Who, with thy saffron wings upon my flowers,
> Diffuseth honey-drops, refreshing showers.

The saffron-crocus was brought to England, it is believed, during the reign of Edward III. Largely cultivated for several centuries in Essex, it gave name to the town of Saffron Walden. Cultivated also close to the metropolis it gave name to Saffron Hill.

Mustard, named several times, and Caraways, mentioned in *2nd Henry the Fourth*, v., 3, have both been in England so long, that if not aborigines, the plants have become perfectly naturalised, and in the Shaksperean age were already common in cultivation. Some think that the caraways of this passage denote a kind of apple, a sub-variety of the nonpareil, called the "caraway-apple," the scene being laid in Shallow's garden:—"Nay, you shall see mine orchard, where in an arbour we will eat a

---

* Æschylus, *Agamemnon*, 239.

last year's pippin of my own graffing, with a dish of caraways, and so forth." But as it is hardly likely that the old gentleman would propose to his guests that they should eat one kind of apple after another, a confection flavoured with the aromatic seeds is most probably the "dish" intended.

To mention such articles of vegetable origin as remain, is necessary only for completeness' sake. They merge, indeed, into manufactures, and no exact line of distinction can be drawn. Rice appears in the *Winter's Tale* passage cited when speaking of raisins (ii., 4). Hemp, as employed for cordage, upon four or five occasions:—

> Upon the hempen tackle ship-boys climbing.
> > *Henry the Fifth*, Act iii., Chorus.

Flax also, about as many times:—

> Shall to my flaming wrath be oil and flax.
> > *2nd Henry the Sixth*, v., 2.

Cork also, upon three occasions; and ebony upon four. Ebony, the quite black heart-wood of different species of Diospyros, large evergreen trees of the eastern tropics, is one of the earliest productions of its kind to be mentioned in history. "The men of Dedan were thy merchants; many isles were the merchandise of thine hand; they brought thee for a present, horns of ivory and ebonies" (Ezek., xxvii., 15). With the ancients this remarkable wood was highly esteemed for inlay cabinet work. Because of the blackness, Ovid uses it to intensify his picture of the abode of Sleep, the bedstock of the god being fabricated of ebony *(Met.* xi., 610).

Shakspere was probably no stranger to it, two of the allusions being literal; the others, as with many of the poets, metaphorical:—

> By heaven, thy love is black as ebony.
> Is ebony like her? O wood divine!
> A wife of such wood were felicity.
>> *Love's Labour's Lost,* iv., 3.
>
> Rouse up revenge from ebon den, with fell Alecto's snake.
>> *2nd Henry the Fourth,* v., 5.

Sugar, wine, vinegar, oil, and their analogues, belong to the idea of the flora even less. They were common commodities of the time, and the references to them acquire their interest on grounds altogether different.

MEDICINES.

Shakspere was no great admirer of the druggists of his day:—

> I do remember an apothecary,—
> And hereabout he dwells,—whom late I noted
> In tattered weeds, with overwhelming brows,
> Culling of simples. Meagre were his looks,
> Sharp misery had worn him to the bones.
> And in his needy shop a tortoise hung,
> An alligator stuffed, and other skins
> Of ill-shaped fishes;—and about his shelves
> A beggarly account of empty boxes,
> Green earthen pots, bladders, and musty seeds.
> Remnants of packthread, and old cakes of roses,
> Were thinly scattered, to make up a show.
> Noting this penury, to myself I said—
> And if a man did need a poison now,
> Whose sale is present death in Mantua,
> Here lives a caitiff wretch would sell it him.
>> *Romeo and Juliet,* v., 1.

The contents of this dismal abode would be chiefly dried herbs. Originally there were no medicines but these, as implied in the very name of the apothecary's profession, "drug" and "druggist" being derived from the Anglo-Saxon *drigan*, to dry. The extension of the name to mineral medicines is comparatively modern. Among the former, in the Shaksperean age, would be all the plants which still bear the botanical epithet *officinalis*, literally, kept in the drug-shops, as Veronica *officinalis*, Althæa *officinalis*, Melilotus *officinalis*. Here, too, we are led to suppose, could be purchased senna, received probably from Egypt; and rhubarb, from western Asia, *viâ* Turkey, whence the old familiar epithet—people being prone to confound the place whence an article is proximately received with the actual and original source,—both drugs mentioned in *Macbeth*, v., iii. Aconite, prepared from the purple monkshood, *Aconitum Napellus*, would also be got there *(2nd Henry the Fourth*, iv., 4). Colocynth, also, mentioned in *Othello*, i., 3, under the name of "coloquintida,"—the inexpressibly bitter pulp of the gourd - like *Cucumis Colocynthis*, indigenous to Palestine, where it supplied the "apples of Sodom." Opium, also, mentioned in *Othello*, iii., 3, under the name of the plant, the renowned "poppy," *Papaver somniferum*, from which it has been prepared for twenty-five centuries.

With the poppy is coupled the name of the other chief opiate of the period, the, if possible, still more renowned mandragora. The passage is that fearful one where

Shakspere's most hideous character, the impersonation
of infernal craftiness, gloats over the successful issue of
his malignity.   Enter Othello—

> IAGO: Look where he comes ! Not poppy, nor mandragora,
> Nor all the drowsy syrups of the world,
> Shall ever med'cine thee to that sweet sleep
> Which thou owed'st* yesterday !—(iii., 3.)

No rest, no repose, ever again for thee, Othello! Opiates
enough to cast a nation into slumber with thee, Othello,
henceforth shall avail not any.   A reflection more
diabolically cruel it is impossible to imagine.   Not content with seeing Othello the victim of his remorseless
villany, Iago rejoices in the thought that he has secured
for him the torture of sleeplessness.   Never again will
he go shares in the benevolent gift which nature allows
to the most forlorn:—

> The innocent sleep,
> Sleep that knits up the ravell'd sleeve of care,
> The death of each day's life; sore labour's bath;
> Balm of hurt minds; great Nature's second course;
> Chief nourisher in life's feast.—*Macbeth*, ii., 2.

Othello stands in the front rank of the Shaksperean
characters whose sorrowful fate moves our souls to profoundest pity.   Magnanimous, ardent, capable of the
sincerest affection, innocent in all his own natural
impulses until so infamously deceived, this last through

---

* Owed'st, *i.e.* owned or had the command of, especially when the
speaker would imply an absolute right to the thing possessed, a
word met with elsewhere, as in the *Two Gentlemen of Verona*—
"That such an ass should owe them."

V

the credulousness which so often belongs to a noble heart—it is the firmness, the resoluteness of his great spirit which at last precipitates the dreadful event. Nothing of its sort is finer in Shakspere than the gradual poisoning of Othello's mind; the little by little growth of the infinite misery; the faltering of the voice as the unhappy man becomes conscious of his shame; with at last, when too horribly convinced, the magnificent simile with which he meets the insidiousness of his fiendish enemy:—

> IAGO: Patience, I say; your mind perhaps may change.
> OTHELLO: Never, Iago. Like to the Pontic sea,
> Whose icy current and compulsive course
> Ne'er feels retiring ebb, but keeps due on
> To the Propontic, and the Hellespont;
> Even so my bloody thoughts, with violent pace,
> Shall ne'er look back, ne'er ebb to humble love,
> Till that a capable and wide revenge
> Swallow them up.—(iii., 3.)

At last, when he is about to slay the poor guileless creature, how sad and tender the wifely trembling. The last dialogue is not so much spoken as breathed between them. The depth of his love for her is still so vast that a very little might make him relent. It is only by summoning a whirlwind of passion that he is enabled to nerve himself for the tragic close.

Mandragora appears again in *Antony and Cleopatra:*—

> CLEOPATRA: Charmian !
> CHARMIAN : Madam.
> CLEOPATRA: Ha, ha !
>   Give me to drink mandragora.
> CHARMIAN : Why, madam?

CLEOPATRA: That I might sleep out this great gap of time,
My Antony is away.

Brilliant, fascinating, vain, artful Cleopatra, one of the
most striking illustrations in the entire series of Shakspere's characters, of his unsearchable mastery in the
superb art of blending diversities, the picture of the
Egyptian queen in this respect surpassing even that of
Beatrice, and the more admirable because so entirely
in accord with unquestioned history—"Give me," she
means, oblivion. Cleopatra and Beatrice, among Shakspere's female characters, like Hamlet and Othello among
the masculine, illustrate also that wonderful power of the
poet, by which, over and above every other, he addresses
and engages the entire mind of his attentive reader. To
have particular faculties addressed is all very well: far
better is it for us that not one shall lie unsolicited; in
this respect, as years roll on, will the inestimable value
of Shakspere as an intellectual lever be found more and
more to consist.

According to Lyte, mandragora was employed in
the cheerful surgery of the sixteenth century as an
anæsthetic:—"The wine wherein the roote of mandrage
hath been stieped or boyled, swageth all paine; wherefore men do geve it, very wel, to such as they intende to
cut, sawe, or burne in any part of their bodies, because
they shall feel no payne" (p. 437). Shakspere would
seem to intend mandragora also in *Hamlet*, i., 5:—

And duller should thou be than the fat weed
That rots itself at ease on Lethe's wharf.

"Fat" is here to be understood as referring to the dulness and sluggishness induced by obesity, which is an image, in turn, of the paralysing effects of the herb.

The plant supplying this famous drug was the *Mandragora officinalis*, one of the Solanaceæ, indigenous to Palestine, especially upon Mount Carmel and near Nazareth, also to Syria and Greece. Sibthorp, who gives a drawing of it in the *Flora Græca*, 232, says that about Athens it is "not rare." In appearance it resembles a primrose, only that the flowers are purplish-white, and are followed by orange-coloured berries, Rachel's *dudha'im*, above-mentioned, in old authors called "love-apples." The root, long-enduring, is very large, fleshy, and forked, so as to present a grotesque image of the human body. Trimmed with the knife, so as to intensify the likeness, it became the equally famous "mandrake" of the middle ages. To find a vegetable production more burdened with superstitions during that period of history would probably be impossible. The ridiculous little images, called "puppettes" and "mammettes," accredited with magical powers, and made the centre of innumerable absurdities, fetched high prices with the simple. The detail, were it worth while, would fill a chapter; fortunately, the whole matter was exhausted long ago by Sir Thos. Browne, in the "Vulgar Errors."

Shakspere himself adverts to these old fancies, of course not accepting, but in deference to the legends of his time:—

Would curses kill, as doth the mandrake's groan,
I would invent as bitter-searching terms,
As curst, as harsh, and horrible to hear.
2nd Henry the Sixth, iii., 2.

And shrieks, like mandrakes', torn out of the earth,
That living mortals, hearing them, go mad.
Romeo and Juliet, iv., 3.

He refers also to the "puppettes," in *2nd Henry the Fourth*, i., 2, and again in iii., 2.

In connection with the mandragora, it may be interest‑ ing to mention that the Greeks gave this plant, for another name, Circæa, in reference to the renowned enchantress. Circæa, as a botanical name, has long been transferred to a pretty and innocent little flower of the woodlands, which thus inherits also the appel‑ lation, in default of the history so unintelligible, of "Enchanters' nightshade."

The introduction, in *Much Ado*, iii., 4, of another famous medicinal plant of the middle ages, the *Carduus Benedictus*, or "Holy Thistle," brings us back to life's merriment. For now we are again beside Beatrice, cheered by the remembrance, as she enters, that all Shakspere's greatest characters have humour in their souls, and that all these greatest, though they can be serious enough when occasion requires, never take their leave without having at times induced a smile, the priceless function of all that is truly good. At the same moment we have sly, shrewd, roguish Margaret, and dear little Hero, her mistress, who never fails in modesty and gentleness. Beatrice is at loving war

with the "young lord of Padua," Benedick. Margaret
has read that "this worthy hearbe," the *Benedictus*,
" was named the blessed thistle for his singular vertues
as well againste poysons, as the pestilent agues and
other perilous diseases of the heart." Where, when,
ever lived the woman who would let slip so golden a
chance! Beatrice says, "I am sick."

MARGARET: Get you some of the distilled Carduus Benedictus,
and lay it to your heart. It is the only thing for a qualm !

BEATRICE: Benedictus! Why Benedictus? You have some
moral in this Benedictus!

MARGARET: Moral! No, by my troth, I have no moral meaning. I meant plain Holy Thistle !

That we may believe or not, as we like. Beatrice,
pretend as she may, knows quite well what Margaret
means.

Chapter Fourteenth.

BOOK AND HEARSAY NAMES.

Full many a glorious morning have I seen
Flatter the mountain-tops with sovereign eye;
Kissing with golden face the meadows green,
Gilding pale streams with heavenly alchemy.

Sonnet xxxiii.

THE mass of all language, even of such portions as on the surface seem purely literal, consists either of metaphors, or of short crystallised similes. Language teems also, as it must needs, with expressions referring to objects and phenomena of which the people who use them can very seldom have personal knowledge, but which objects no one ever hesitates to talk of as if familiar. The whirlpool and the volcano, crocodiles, sharks, and scorpions, are as much a part of the popular vocabulary as trees and oxen.

The poets' language, most particularly, is crowded with such expressions, many of them primæval. They are employed, however, in very different ways. The inferior man simply copies; he is only a scribe: the master uses them as prisms, and by this we discover him. When Shakspere adopts metaphors from the ancients, we get lamps, as of old; but, as in the Arabian tale, they are "new" ones in exchange for the exhausted. His repetitions never carry the look of the second-hand; his echoes are as charming as the originalities. Possessed of the sweet aptitude for observing personally, first and finest of the fine arts, without which there can be no true greatness, he sees everything that his predecessors saw, and sees it with Shaksperean eyes besides. He knew better, moreover, than to adventure upon uncertain ground, simply for the sake of being novel. Better often than to be very new is it to be very old, stepping forth like one of the goddesses in Homer.

The trees unknown to Shakspere as living objects, but which he introduces upon hearsay, are the palm, the cedar, the pine, the laurel or bay, the myrtle, and the olive. Mention is made also of cypress-wood and of Balm-of-Gilead, neither of which articles did he ever see. One or two of the trees may have been seen by him in some garden when he wrote. His references to them were founded, nevertheless, upon simple book-knowledge.

Of Balm, Balsam, or Balsamum, meaning Balm-of-Gilead, the sevenfold famous medicament of Scripture, little more, even now, is exactly known than that it is yielded by certain trees of the order Amyridaceæ, growing in Abyssinia, and in the land of mystery, ancient Arabia. Fragrant, and, when new, semi-fluid, with keeping it becomes resinous and golden yellow, and is at all times so admirable for physicians' purposes, that no wonder it took the place in language it has held for three thousand years, or that of a metaphor for kindliest soothing and consolation:—

'Twas whispered balm; 'twas sunshine spoken!

The trees are adverted to at the end of *Othello:*—

Drop tears as fast as the Arabian trees
Their medicinal gum.—(v., 2.)

The medicament appears, metaphorically, upon seven or eight occasions:—

My pity hath been balm to heal their wounds.
3rd Henry the Sixth, iv., 8.

Balm of your age,
Most best, most dearest.—*King Lear,* i., 1.

Also upon three or four in the sense of the anointing oil used in coronations,—

Not all the water in the rude, rough sea
Can wash the balm from an anointed king.
Richard the Second, iii., 2.

The Cypress.—The cypress had been introduced from the Levant long before the time of Elizabeth. The very remarkable spire-like figure, then without a parallel in England, and the peculiar foliage, would hardly fail to arrest the poet's attention, did he meet with it. Still, in the Shaksperean references to this famous tree we have only ancient literature renewed, though perhaps indirectly, through the Italian poets. In oriental countries the cypress had been a favourite from very early times for planting near tombs, whence Spenser's epithet, in the *Faëry Queene*, "the cypress funereal," an association which at once explains the line in *2nd Henry the Sixth*, iii., 2, when Suffolk, execrating his enemies, wishes

> Their sweetest shade a grove of cypress-trees.

When, in the *Taming of the Shrew*, ii., 1, Grumio says,

> In ivory coffers I have stuffed my crowns,
> In cypress chests my arras,

he refers to the very ancient reputation of cypress-wood as something imperishable, therefore suited for the protection of valuables.*

The Laurel or Bay-tree.—These are two names for the same plant, the celebrated *Laurus nobilis*, the leaves of which, when rubbed or bruised, give out the odour of cinnamon. Familiar to the ancients, the renown begins, as every one knows, with the fable of the nymph whose name was transferred to it, Daphne, the coy and beautiful maid beloved of Apollo. The chase across the meads, the flowing robe and the "careless locks" tossed backwards by the wind, the failing strength, the sense of the god's shadow at her feet, the mournful look,

---

* Cypress, or Cyprus, was the name also, in the Shaksperean age, of some kind of woven fabric, black, it would seem, from the reference in Autolycus' song, in the *Winter's Tale*, iv., 3; and used particularly for mourning, as indicated in the song in *Twelfth Night*, ii., 4.

the piteous cry, "Oh, father, help me!"—these count
with the things that cannot perish. Upon these in turn
came the consecration of the tree to genius and victory,
and in echo of this it is that we find the laurel in Shakspere. Possibly, as above said, he may have been shown
it in some curious garden, but the allusions are in
every instance imitative. Such are "laurelled victory"
in *Antony and Cleopatra*, i., 3, the "laurel crown"
adjudged to Warwick,

> As likely to be blest in peace and war,
>
> *3rd Henry the Sixth*, iv., 6,

and that one contained in the beautiful line in *Troilus
and Cressida*, i., 2,

> Prerogative of age, crowns, sceptres, laurels.

"Bay," the English form of the French *baie*, a contraction
in turn of the Latin *bacca*, berries, the purple and
aromatic fruit, which in the middle ages was much
esteemed, appears in *Richard the Second*, ii., 4:—

> 'Tis thought the king is dead, we will not stay,
> The bay-trees in our country all are withered.

The expression is of course to be understood as a purely
figurative one. Perhaps, as the laurel is the emblem of
nobleness, it may be a metaphor for "all that was great
and good in the country has perished." Richard,
"weak in adversity, reckless in prosperity," had by his
favouritism done mischief that it took many long years
to repair, so that the metaphor would suit. More
probably, being associated with other prodigies,

> These signs forerun the death or fall of kings,

the reference, as said on page 12, may be to some curious superstition, now forgotten. For the flourishing of the bay was a presage of good, and a guarantee of local safety:—"Neyther falling sycknes, neyther devyll, wyl infest or hurt one in that place whereas a bay-tree is. The Romaynes called it the plant of the good angell."* Withering and dying, without apparent physical cause, would, on the converse, necessarily induce conditions exactly opposite.

The pleasant odour of the bay recommended it in olden times for garnishing dainty and stylish dishes on the occasion of banquets. Hence the figurative phrase in *Pericles*, iv., 6, shamelessly spiteful.

The Myrtle.—Nature holds few shrubs that in comeliness and neatness rival the myrtle. The shining green of the perennial foliage, the sweet and uncontested delicacy of the white bloom, just touched with purple, and the pleasant odour, with by and by agreeable fruit, very early declared it a fitting emblem, as in Scripture, of the Christian affections. In secular poetry it quite naturally passed on to the signification of love in general. The Greek and Latin poets adduce it in this connection, in very many places; Shakspere follows suit in the *Passionate Pilgrim*, and in *Venus and Adonis*. In *Antony and Cleopatra*, iii., 10, it enters into a fine comparison:—

> I was of late as petty to his ends,
> As is the morn-dew on the myrtle leaf
> To his grand sea.

---

* Lupton. *Notable Things;* the Syxt Booke.

In *Measure for Measure*, ii., 2, there is another, in which a metaphor is simultaneously involved:—

> Merciful heaven,
> Thou rather, with thy sharp and sulphurous bolt,
> Split'st the unwedgeable and gnarlèd oak,
> Than the soft myrtle.

The Palm.—Excepting in one instance, Shakspere's "palm" is the immemorial tree of Scripture and of classical poetry, the source of dates, and the beautiful symbol, in art, of Palestine and of the Orient in general. Recommended for symbolical use by the columnar form of the lofty and unbranched stem, the steady aspiration, the majestic arching crown of huge evergreen leaves, and its munificence as a fruit-bearer, the palm became the synonym, at a very early period, of nobleness, virtue, victory, and prosperity. Once established as a metaphor, a word like "palm," possessed as it is of a certain happy brevity, a certain pleasant ring or resonance, a certain air of the picturesque, a welcome adaptedness to supply a rhyme or sustain a metre, lives for ever. Men ask no questions about the tree; they know what they mean when they "bear the palm," and "carry the palm," and that is enough. Shakspere certainly never saw a palm; he simply adopts the metaphor, and this always in pleasant fashion, as in *Julius Cæsar*, i., 2; *Coriolanus*, v., 3; *Timon of Athens*, v., 1; and *Hamlet*, v., 2.

When, in *Henry the Eighth*, iv., 2, the dying queen has her beautiful vision of the angels, "clad in white robes, wearing on their heads garlands of bays, and having palms in their hands," the latter part of the imagery, it hardly needs the saying, is adopted from the Apocalypse, the language of which is explained, in turn, by the literal history contained in the fourth Gospel. "The branches of palms" mentioned by St. John, xii., 13, must not be understood, however, as palm-leaves pure and simple, and newly cut for the occasion. The Jewish custom was to prepare these leaves for ceremonial use by trimming them, then binding with myrtle upon the right, and citron upon the left. They were carefully laid by for solemn occasions, and were now fetched out to do honour to the entry into the holy city.

That the palm should become identified with Palestine, and with Jerusalem in particular, was quite consistent. Hence the name of *palmifer*, shortened into "palmer," given to pilgrims to the Holy Sepulchre who had honourably performed the journey, and had now returned, or were returning, homewards, carrying a leaf, or a portion of one, as attestation. The palmers of the middle ages must not be confounded with pilgrims in general. The pilgrim—*peregrinus*—sets out to any one of many different places, according to the object in view; the "palmer's" goal is purely the Holy Sepulchre. The origin and history of the pilgrims, including the palmers, constitutes one of the most interesting chapters in the social life of Europe during the middle ages. The Crusades may be regarded as a portion of it, expressed upon a scale of consummate magnitude. Little by little it leads on to pilgrimages to shrines and famous sanctuaries; the "Canterbury Tales" give an instance of the actuating spirit, only with a dash of caricature. Both pilgrims and palmers are several times mentioned by Shakspere, as in *All's Well*, iii., 5, when Helena makes her appearance, dressed in the style that declares her purpose :—

> God save you, pilgrim! Whither are you bound?
> HELENA: Where do the palmers lodge, I do beseech you?

The exceptional Shaksperean reference to the palm— the date-palm *not* being intended—occurs in *As You Like It*, iii., 2, when Rosalind, rejoicing over her verses, exclaims, "See what I have found upon a palm-tree!" She has sometimes been supposed to mean the palm-willow, *Salix Capræa*, the long and pliant twigs of which, covered with opening catkins, have been used in England for many ages, on Palm-sundays, as a substitute for the branches of the Palestine plant. But there is no need of conjecture of any kind, the signification and the source whence the word was in the present instance derived being plain. *As You Like It* had for its antetype a novelette by Thomas Lodge, originally published in 1590, under the name of "Rosalynde, Euphues' Golden Legacie." "Rosalynde," in turn, had for its antetype the old poem of "Gamelyn,"—authorship uncertain—so hard is it to get at the actual beginning of anything. Often graceful and musical, but full of quaint conceits, the natural history of Lodge's tale is of the oddest,—a circumstance not surprising when we note what is very evident, that it was taken at random from another

predecessor, the celebrated Bartholomew Glanville.
Glanville's book, styled *De Proprietatibus Rerum*, is a
very curious one, comprising a Herbal, a Gazetteer, a
Guide to Domestic Medicine, and some chapters upon
music. Composed about 1360, it was translated into
English by John de Trevisa, and originally printed by
Wynkyn de Worde, about 1494, from which time it
continued, for quite one hundred and fifty years, one of
the most popular of the volumes then read. Shakspere,
as well as Lodge, no doubt was well acquainted with it,
the latter picking out for his "Rosalynde," the "mir-tree,"
the "cypresse-tree," "lymons and citrons," the "fig," a
"hungry lyon," and finally the "palm." The tree meant by
Glanville is the famous "phœnix" of the Græco-Egyptian
myth, reported to flower only once, perishing immediately afterwards, though at once succeeded by another.
Hence, in Lodge, "Thou art old, father Adam, and
thy hair waxes white; the palme-tree is already full of
bloomes, and in the furrows of thy face appear the
Kalenders of death." Founded, like all other myths,
upon a fact in nature, the curious old tale was without
doubt a mistaken history of the talipat-palm, *Corypha
umbraculifera*, of Ceylon; with the merit also of being
designed to illustrate the grandest of physiological
realities, — the Rejuvenescence of nature. Rosalind's
"palm," adopted by Shakspere from Lodge, is thus
nothing more than the Egyptian phœnix. The fable of
the tree got confounded, as well known, through similarity
of name, with that of the bird. Allusion to the latter

is made by Sebastian in the *Tempest*, iii., 3, "One tree, the phœnix' throne." It is mentioned also in *Antony and Cleopatra*, iii., 2, and in *Henry the Eighth*, v., 4.

When, in *Hamlet*, Horatio says,

> In the most high and *palmy* state of Rome,

it is questionable if the reference be not to something very different. The antlers of deer, fallow-deer, at all events, when full grown, are called "palms," and in their maturity are said to be "palmate." The word is employed in this sense in Chapman's translation of the *Iliad*, iv., 124. Pope, Cowper, and, in our own day, Lord Derby, employ it in the same manner; and Shakspere, to whom the "dappled fools," were beloved familiars, would be quite as likely to cite their perfection as an emblem of imperial power, as to fall back upon the stereotyped image of a tree he had never seen.

The Olive.—In two places in *As You Like It* the olive-tree is mentioned as an element of living nature.

> If you will know my house,
> 'Tis at the tuft of olives here hard by (iii., 5);

then in iv., 3,—

> A sheep-cote fenced about with olive-trees.

In both, however, the name appears after the same manner as Rosalind's "palm;" by adoption, that is to say, from the novelette which supplied the original idea of this beautiful play. Matchless Arden, though Shakspere gives us, in its leafy coverts, his own beloved Warwickshire, is Fairyland all the same—Fairyland of

w

the best and truest kind. It is the play in respect of which, above all others, we are least entitled to press for literal consistency in the natural history. If we are, well then good-bye alike to the olive, to the "green and gilded snake," to the lioness that "lay couching, head on ground,"

> When that the sleeping man should stir;—

good-bye, in a word, to Rosalind.

Elsewhere the allusions to this famous plant correspond with those to the laurel. They relate, in every instance, to the very ancient symbolical or representative use of the olive, which in Scripture, as well as in classical verse, is constantly put for charity, peace, and good - will. This, though he never saw an olive, Shakspere knew quite well. In *Twelfth Night*, "I bring," says Viola, "no overture of war, no taxation of homage. I hold the olive in my hand; my words are as full of peace as matter" (i., 5). So in *2nd Henry the Fourth*, iv., 4,—

> There is not now a rebel's sword unsheathed,
> But peace puts forth her olive everywhere.

Similar words occur in *Antony and Cleopatra*, iv., 6; *Timon of Athens*, v., 5; *3rd Henry the Sixth*, iv., 6; and in Sonnet cvii.

The Pine. — The magnificent group of conifers so well distinguished by the tesselated cones and long acicular leaves, includes as its British representative that one commonly called the "Scotch fir," the *Pinus sylvestris* of the botanists. Whether Shakspere knew

much, or indeed anything, of this tree, is very doubtful.
Not a companion of the oak, preferring bleak mountains
to the plain and the sheltered valley, even if it existed
in Warwickshire when the plays were written, it would
not form a feature in the landscape. Neither would any
other conifer be conspicuous; the Norway spruce alone,
Abies excelsa, had been introduced, and this was not a
common plant. When, accordingly, we come upon
references to the "pine," the name must be understood
as one of those adopted from the classical poets, who
abound with allusions to the beautiful forms indigenous
to southern Europe, the *Pinus Pinaster*, the *Pinus Picea*,
the *Pinus Laricio*, and perhaps others, but without
discrimination of species. One or two of the references
are purely literal:—

> Behind the tuft of pines I met them.—*Winter's Tale*, ii., 1.

One or two others are figurative,—

> Thus droops this lofty pine, and hangs his sprays.
> *2nd Henry the Sixth*, ii., 3.

The finest of all belong to the former class, as in that
most splendid picture, in its kind not excelled by any
poet ancient or modern, where the rising of the sun is
spoken of:—

> But when from under this terrestrial ball
> He fires the proud tops of the eastern pines.
> *Richard the Second*, iii., 2.

The pines, that is to say, upon the mountain-crest, are
the first to be touched by the light, all golden, which at
sunset they are the last to surrender. One cannot but

think of the "great god Pan," whose name denotes universal nature, and whose brows were represented as wreathed with pine-foliage, the poet's sweet way of saying that the mountain-outlines of the world are fringed with this glorious tree. Pine-*wood* (possibly that of the *Pinus sylvestris*) is referred to in *Troilus and Cressida*, i., 3:—

> As knots, by the cònflux of meeting sap,
> Infect the sound pine, and divert his grain,
> Tortive and errant, from his course of growth.

The knots induce, nevertheless, a rich and beautiful variegation, plain enough when a slice is polished,—

> Checks and disasters
> Grow in the veins of actions highest reared,—

the sense of the whole being identical with the spirit of *As You Like It*, and with the idea set forth in so many other places, that men of noble heart learn in conflict with vicissitude, not so much how weak they are, but how strong.

Pine-wood, charged with resin, is highly inflammable. This is why Ceres, searching for her lost daughter, the whirled-away in "Dis's wagon," from which were dropped the immortal daffodils, restless even by night, kindled "pines" in the flame of Ætna to serve as torches. How beautiful, too, the citation of the pine when forsaken Medea, describing the fervour and rapidity with which her affection had been kindled for Jason, compares it to the sudden and brilliant ignition of a "pine-brand," when lighted at a sacrifice to the gods. Is it permissible to think that it may have been wood of the same tree which was used for Prospero's hearth? Enough if the

thought transforms into the sound of the voice of charmed
Miranda, to whom self-sacrifice is heaven. These few
words show that the sweet "eye-wedlock" is complete:—

> Alas now! pray you,
> Work not so hard! I would the lightning had
> Burnt up these logs, that you are enjoined to pile;
> Pray set it down, and rest you: when this burns
> 'Twill weep for having wearied you.
> FERDINAND: O most dear mistress,
> The sun will set before I shall discharge
> What I must strive to do!
> MIRANDA: If you'll sit down,
> I'll bear your logs the while. Pray give me that,
> I'll carry it to the pile.—*Tempest*, iii., 1.

Coniferous trees supplied also, in ancient times, the
fragrant wood burned as incense in sacred rituals. The
"wood-of-sacrifice" mentioned by Homer in his description of the enchanted island of friendly Calypso reappears
in the Apocalypse as "thyine-wood." In Shakspere
we have something of the same sort in the *Taming of
the Shrew*:—

> Carry him gently to my fairest chamber,
> And burn sweet wood to make the lodging sweet.
> *Induction.*

The Cedar.—By this name is* always understood in
literature, the immemorial cedar-of-Lebanon, *Cedrus
Libani*, so often employed in Scripture as an emblem of
kingly stateliness and magnificence. It was the scriptural
use which furnished Shakspere with his ideas of the tree,
no mention being made of it by the classical authors, to
whom, indeed, the *Libani* would seem to have been

unknown. As an object of nature, there is nothing perhaps which in the aggregate is more imposing. The qualities, as with all other truly great and glorious things, are such as not only catch the eye, but retain it. The stature, the girth, the measured symmetry, the remarkable and unique spread of the huge boughs, which strike out in such a way as to form distinct horizontal stages or terraces, are separately and independently enough to confer nobleness; and in addition there is the prophet's "shadowy shroud," no tree being more remarkable for closely woven density of dark leafage, never lost. Shakspere introduces it upon twelve different occasions.

> The sun ariseth in his majesty,
> Who doth the world so gloriously behold,
> That cedar-tops and hills seem burnished gold.
> *Venus and Adonis.*

> As on a mountain-top the cedar shows,
> That keeps his leaves in spite of any storm.
> *2nd Henry the Sixth*, v., 1.

> He shall flourish,
> And like a mountain-cedar reach his branches
> To all the plains about him.—*Henry the Eighth*, v., 4.

Sometimes it is used as an image of grandeur:—

> The lofty cedar, royal Cymbeline,
> Personates thee.—*Cymbeline*, v., 5.

> Marcus, we are but shrubs; no cedars we.
> *Titus Andronicus*, iv., 3.

> Thus yields the cedar to the axe's edge,
> Whose arms gave shelter to the princely eagle.
> *3rd Henry the Sixth*, v., 2.

Other examples, various in their kind, occur in the *Tempest,* v., 1; *Love's Labour's Lost,* iv., 5; *Richard the Third,* i., 3; *Coriolanus,* v., 3; *Cymbeline,* v., 4 (repeated in v., 5); and the *Rape of Lucrece.*

No cedars existed in England during Shakspere's lifetime. The first individuals appear to have been raised from seed by the celebrated Evelyn, author of the *Sylva,* some time about 1680. The Warwick Castle cedars, though popularly assigned to the Norman period, are thus not yet two centuries old. As a matter of fact they were planted only about 1740, by the grandfather of the present earl. They would then probably be vigorous saplings of some few years' growth; and to the near neighbourhood of the river, the shelter, and the soil, may be ascribed the grand proportions they now exhibit.

No slight privilege is it when visiting Shakspere's home to find, within distance so easy, a feudal relic so grand, and upon which his eyes must often have rested;—and beside it, trees which, though he never beheld, were in their storied kind so richly present to his imagination. These Warwick Castle cedars are in a measure like himself, towers of grandeur reared in the past, more majestic every day, and that will be stately and commanding as long as it is possible for such things to endure. They reckon with the things which awaken gratitude in its best forms. It is counted fine to raise a splendid pile of stone or marble, to paint a sublime picture, to compose a brilliant opera. Perhaps the man who multiplies trees of glorious sort, especially when, like these cedars, they compare with epitaphs, achieves in his day quite as genuinely as good an end.   He makes the world richer than he found it, a good that any man may be proud to accomplish. He can hope, in any case, when the sun is sinking in the west, that though his name may slide away and be forgotten, the work he has done, or has essayed to do, so that it has been honest and faithful, may not have been altogether in vain; that it will give pleasure to a thousand hearts yet unborn, and inspire a thousand more to go and do likewise.   I would rather be able to reflect in my old age that I had been the originator of a hundred oaks and cedars that in days to come shall help to make my country glad and beautiful, than have it recorded of me, simply and exhaustively, that my will was "proved under a million," and leave no memorial besides.

INDEX.